D1328538

Gravity's Ghost

Gravity's Ghost

Scientific Discovery in the Twenty-first Century

HARRY COLLINS

The University of Chicago Press
Chicago and London

HARRY COLLINS is distinguished research professor of sociology and director of the Centre for the Study of Knowledge, Expertise, and Science at Cardiff University. He is the author of *Tacit and Explicit Knowledge* and *Gravity's Shadow* and coauthor of *Rethinking Expertise*, all published by the University of Chicago Press.

The University of Chicago Press, Chicago 60637
The University of Chicago Press, Ltd., London

Printed in the United States of America

20 19 18 17 16 15 14 13 12 11   1 2 3 4 5

ISBN-13: 978-0-226-11356-2 (cloth)
ISBN-10: 0-226-11356-6 (cloth)

Library of Congress Cataloging-in-Publication Data

Collins, H. M. (Harry M.), 1943–
Gravity's ghost : scientific discovery in the twenty-first century / Harry Collins.
p. cm.
Includes bibliographical references and index.
ISBN-13: 978-0-226-11356-2 (cloth : alk. paper)
ISBN-10: 0-226-11356-6 (cloth : alk. paper) 1. Gravitational waves—Research—History. 2. Gravitational waves—Experiments. 3. Science—Social aspects.
I. Title.
QC179 .C646 2011
530.11—dc22

2010013704

♾ The paper used in this publication meets the minimum requirements of the American National Standard for Information Sciences—Permanence of Paper for Printed Library Materials, ANSI Z39.48-1992.

CONTENTS

INTRODUCTION

I begin to write this volume right now at 2:10 p.m. U.S. West Coast time on 19 March 2009. That's the first sentence. I am sitting in Los Angeles airport with an hour or two before my flight back to the UK. I have just come from the small California township of Arcadia, where I have been attending a meeting of the "LIGO Scientific Collaboration and Virgo." The Collaboration is six or seven hundred strong, and it has been trying to detect gravitational waves using apparatuses costing hundreds of millions of dollars. These giant machines have been "on air" for a couple of years, and a lot of data has been collected which has only just been analysed; it might or might not contain intimations of a gravitational wave. The high point of the meeting was the "opening of the envelope"—an event that has kept me and most of the gravitational wave community on tenterhooks in the eighteen months since the autumn of 2007.

The "envelope" held the secret of the "blind injections." The blind injections were the possible introduction of fake signals into the data stream of LIGO—the Laser Interferometer Gravitational-Wave Observatory. The idea was to see if the physicists could find them. Only the two people who had injected the signals knew the secret before this meeting.[1] Their brief was to inject fake

1. Or so I thought until Jay Marx, the executive director of LIGO, told me (private communication, October 2009), that he had known the secret from the beginning and had been keeping a "poker face" through the whole eighteen months—a remarkable performance.

gravitational wave signals according to a randomized code. Both the shape and strength of the signals and the number of signals injected was to be decided at random. One possibility was that no signal at all had been injected. It could also be that one, two, or even three signals had been injected. On this depended whether anything suggestive in the data was a blind injection and whether any blind injection had given rise to anything suggestive. Last Monday, in that tense meeting, I and the rest of the community were told the truth; today is Thursday. Before you finish reading this book you too will know what was in the envelope. To get the greatest benefit from the book, however, I would suggest, don't flip to the end—follow events as I and the physicists lived them; read the book in the spirit of a "whatwosit," the physics version of a "whodunit."[2]

The story recounted in these pages is part of the history of gravitational wave detection that I started to document in 1972.[3] Since a second period of very intense involvement began in 1994, I have been given more and more access to the field. I now work from a privileged and possibly unique position—an outsider, with no controls on what I disclose except those of good manners and good sense, who is privy to the inner discussions of a live scientific group as they struggle to make a discovery.

Only since the mid-1990s, with the invention of the unobtrusive digital voice-recorder, has it been possible to tell a story with what the actors said in real time playing the large role it does here. With such a device I can sit quietly in the corner of the rooms in which the events I report take place, typing notes into my laptop computer and recording anything that anyone says that sounds interesting. Of course, that I am allowed to do this as a matter of course is also something worthy of remark. It has been a long and slow matter of building trust and colleagueship with the members of the gravitational wave community, something which has been not an onerous task for me but part of my reward for doing my job. Why did the scientists ever allow me to get started? After all, some of what I write might embarrass them. It is because they believe in the academic enterprise and know this is the right thing to allow, even if it does make for a less comfortable life. On the upside for the scientists, we can be sure of one thing: any group that is prepared to allow an outsider like me to listen in on their innermost

2. I respectfully ask reviewers not to give away the secret either.

3. The period from the start to around 2003 is described in my book, *Gravity's Shadow* (2004). See also www.cf.ac.uk/socsi/gravwave.

discussions is a group you can trust. What is strange is that I am probably the only person who is currently doing work of this kind.[4]

Over the years I have come to love gravitational wave detection physics and the people in it. As a sociologist of science I have investigated quite a few fields but chose to do my career-long fieldwork study following gravitational waves because I felt more at home in this science that in any of the others I looked at. The task the physicists have set themselves is nearly impossible, and it will, and then only with luck, take a lifetime to complete. Expecting little financial reward, they spend their existence encountering endless frustrations and disappointments in the hope of gaining a miniscule increase in understanding of how the world works; I find myself happy in such company. At times of despair and encounters with stupidity, the example of the gravitational wave physics community has rekindled my faith in the worlds of both science and social science. Ironically, given what is in this book, it has helped me to continue to believe that a high standard in the search for truth is better than academic *realpolitik*, both in theory and in life. In what I call the "Envoi," I try to elevate this point into a political philosophy, arguing that science done with real integrity can provide a model for how we should live and how we should judge.

The irony is that what I describe may be scientists trying too hard to achieve perfection. In the twenty-first century it may be better to allow the imperfection of *the best that can be done*—which gravitational wave detection physics assuredly is—to be revealed, not disguised. The model that has led, or perhaps, misled, the philosophical understanding of the nature of science for too long is Newtonian physics, along with its successors, relativity, quantum physics, high energy physics, and so forth. The gold standard for these sciences is exact quantitative prediction, triumphantly confirmed in more recent years via statements of high levels of statistical significance. Retrospective accounts of these triumphs have given us an unsupportable model of how the world can be known.

Two things have gone wrong. The first thing is that the domain to which the Newtonian model applies, though it takes up a huge proportion of that part of our imaginations that is devoted to science, is a tiny and unrepresentative corner of the scientific enterprise. Nearly all science is a mess—think about long-term weather forecasting, the science of climate,

4. Try to imagine how different certain commercially driven sciences might be if they let people like me listen to their inner deliberations. Wouldn't it be better if there were this kind of record of more major scientific enterprises, both taxpayer-funded and private?

the science of human behavior, economics, and so on and so on. The trick of the sciences of the very large, and the sciences of the very small, is that nothing much happens in outer space and nothing much happens in inner space; get away from the Earth and there is not much there; get down to where it is all spaces between subatomic particles and there is not much there either. That is why the sciences of astronomy, astrophysics, and cosmology, and of quantum and high-energy physics, are so simple, and that is why they can more easily appear to match the idealized model. Down/up here, where most of us live from day to day, there is so much going on that it is almost impossible to make a secure prediction. So the Newtonian model is unrepresentative of science in a statistical sense and still less representative of the sciences of pressing concern to the citizen.

The second thing that has gone wrong is that the Newtonian model is not even a correct description of itself. The triumphal accounts are either retrospective, or refer to sciences that are so well established that they have made all their mistakes and become technologically secure—not reaching, but perhaps reaching toward, the reliability of your fridge or your car. The revealing science—the science which more readily shows us how humans wrest their understanding from a recalcitrant Nature—is pioneering science, where things are being done for the first time and mistake after mistake is being made. Gravitational wave detection physics is a true science in this respect. Here things are being done for the first time.

The contrast between the frontier sciences and the technologically well-developed sciences is a main lever of analysis in this book, and it may be that the sociologist's deliberately distanced perspective helps to bring it out. On the other hand, the views that emerge from the sociological perspective, at least insofar as they bear on the dilemmas of the science itself rather than its role in society, are not dissimilar from those of some of the members of the scientific collaboration being studied. The sociological contribution is, perhaps, simply to set out the arguments in a systematic way and relate them to wider issues.

Some of these sentiments might grate on those brought up in the tradition of science and technology studies, or "science studies," as it has been practiced since the early 1970s. The prevailing motif of the field has been the "deconstruction" of the idealised model of science. That I could argue above that the idealised model is not even a good description of the Newtonian sciences and their counterparts is a result of the new understanding that has come with this movement—a movement in which I have been involved from the beginning. The sentiments expressed above have to be understood in the light of what has been called the "three wave" model of science studies.[5] The First Wave took science to be the preeminent form of knowledge-making, and the job of philosophy of science was to tease out its logic and that of the social study of science to work out how society could best nurture it. The Second Wave used a variety of skeptical tools—from philosophical analysis to detailed empirical studies of the day-to-day life of science—to show that the predominant model of science was wrong and that the examples of scientific work used to support it were oversimplified. For example, the Michelson-Morley experiment was regularly described as having shown, in 1887, the speed of light to be a constant, whereas it took fifty years of dispute before scientists agreed that its results were empirically sound.[6] Most of this book is Wave Two science studies: it is going to reveal how difficult it is be sure of what you are finding out even in physics.

Wave Two shows, with the quasi-logical inevitably of the application of skepticism, that science has no philosophically or practically demonstrable special warrant when it comes to knowledge-making. The recently proposed Wave Three of science studies accepts this but argues that technically based decisions still have to be made in a modern society. Wave Three therefore seeks an alternative way of establishing the value of the science-driven thinking that we are almost certainly going to put at the heart of our technical judgements. The proposed alternative is the analysis of expertise.[7] Wave Three makes it explicit that in spite of the logic of Wave Two, which shows how sciences claim to true knowledge can be "deconstructed," science is still the best thing we have where knowledge about the natural world is concerned.[8] Here, the processes of science are unapologetically spoken of as the most valuable models for the making of technological knowledge, even though this cannot be proved to be the case by detailed description or logical analysis. Gravitational wave physics, in spite of the fact, or perhaps because of the fact, that what it counts as a finding has to be wrested from the fog of uncertainty by human technical judgment, is an example of the best that humans can do and should do.

5. Collins and Evans 2002, 2007.

6. Collins and Pinch 1998.

7. Collins and Evans 2007.

8. Another thing discussed in this book that does not mesh easily with contemporary science studies is that gravitational wave physics can be contrasted with other sciences, such as high-energy physics, which can be treated as, in some sense, more perfectible. Even though it has been shown that even in the heartland of science there is no perfection to be found, the contrast between the two kinds of science remains a useful device, so long as it is understood in the spirit of saying that it is more certain that the Sun will rise tomorrow morning than that tomorrow's weather will be like today's, even though both are subject to the problem of induction.

1 Gravitational Wave Detection

A Brief History of Gravitational Wave Detection

In 1993 the Nobel prize for physics was awarded for the observation, over many years, of the slow decay of the orbit of a binary star system and the inference that the decay was consistent with the emission of gravitational waves. Here, however, we are concerned with the detection of gravitational waves as a result of their "direct" influence on terrestrial detectors rather than on stars. The smart money says that the first uncontested direct detection will happen six to ten years from now, almost exactly fifty years since Joseph Weber, the field's pioneer, first said he had seen them. Joe Weber's claim was not uncontested. It was one of some half-dozen contested claims to have seen the waves made since the late 1960s. All of these have been consigned, by the large majority of the physics community, to the category of "mistake."[1] The rejection of these results by the balance of the gravitational wave

1. This volume, though it can be read independently, is the second in a series that should have (at least) three volumes—the first being *Gravity's Shadow*, my account of the sociological history of the attempt to detect gravitational waves from its beginning. This includes the remarkable story of the contested claims and the shift from small science to the new, big-science, technique of interferometry. *Gravity's Shadow* ends around 2003. The third volume should be about the uncontested claims to come. The current book has few bibliographical references, and those who feel the need to go back to the original sources pertaining to the history of gravitational wave detection, or the science studies background of what is argued here, should refer to *Gravity's Shadow*.

community was often ferocious, driven by the sense of shame at the field's reputation for unreliability, or flakiness, in the eyes of outsiders. Newcomers to the enterprise also had to justify spending hundreds of millions of dollars on the much larger instruments—the giant "interferometers"—that they felt would finally be able to make a sound detection and atone for past mistakes; if the old cheap technology really could see the waves, then there would be no need for the new, so the credibility of the old cheap technology had to be destroyed.

The proponents of the old technology fiercely resisted the destruction of their project, which caused both sides to dig themselves into polarized positions.[2] The consequence was that for decades the creative energy of most interferometer scientists was directed at finding flaws; the principle activity had become showing how this or that putative signal in either their rivals' or, subsequently, in their own detectors was really just noise. This is the problem of the negative mindset that is a central feature of what is to follow. In the meeting in Arcadia that bitter history stalked the corridors with an almost physical presence.

Weber and the Bars

Joe Weber was a physicist at the University of Maryland. In the 1950s he began to think about how he might detect the gravitational waves predicted by Einstein's theories. Gravitational waves are ripples in space-time that are caused by rapid changes in the position of masses, but they are so weak that only cosmic catastrophes such as the explosion or collision of stars or black holes can give rise to enough of the radiation to be even conceivably detectable on the surface of the Earth. It would take a great leap of the imagination, a genius for experiment, and a heroic foolhardiness to try it. Weber was equipped with the right qualities, and he built a series of ever more sensitive detectors; by the end of the 1960s, he began to claim he was seeing the waves.

Weber's design was based on the idea that ripples in space-time could be sensed by the vibrations they caused in a mass of metal. He built cylinders of aluminum alloy weighing a couple of tons or so and designed to resonate—to ring like a bell at around the frequency of waves that might plausibly be emitted by a source in the heavens. Every calculation of the energy in such waves and the way they would interact with Weber's detec-

2. More of this battle will be described in the odd-numbered chapters; these provide background and commentary. The even numbered chapters tell the story.

tors implied that he did not have a hope, and when he started he did not think he had a hope either. But he went ahead anyway.

Weber insulated the cylinders from all the forces one could think of, but to see a wave it was necessary to detect changes in the length of the cylinders of the order of 10^{-15} m, the diameter of an atomic nucleus, or even less. Vibrations of this size, however, are continually present in the metal anyway, no matter how carefully it is insulated. Crucially, Weber built two of the cylindrical devices and separated them by a thousand miles or so. Then he compared the vibrations in the two cylinders. The idea was that, if there was a coincident pulse in both detectors, only something like gravitational waves, coming from a long way away, could cause it.

Since both of the cylinders would suffer from random vibrations, there were bound to be coincident pulses every now and again just as a result of chance. But Weber used a very clever method of analysis. He used something called the "delay histogram," which is nowadays referred to as the method of "time slides" or "time shifts"—a method that is still at the heart of gravitational wave detection forty years on and that will be at the heart of the method for the foreseeable future. Imagine the output of the detector drawn on a steadily unwinding strip of paper, as in those machines that record the changing temperature over the course of a day, but, in this case, sensing vibrations microsecond by microsecond; it will be a wiggly line with various larger pulses impressed upon it. One takes the strip from one detector and lays it alongside the strip from the other. Then one can look at the two wiggly lines and note when the large pulses are in coincidence. Those *coincidences* might be caused by a common outside disturbance such as a gravitational wave, or they might be just a random concurrence of *noise* in the two detectors. Here comes the clever bit: one slides one of the strips along a bit and makes a second comparison of the large pulses. Since the two strips no longer correspond in time, any coincidences found can only be due to chance. By repeating this process a number of times, with a series of different time slides one can build up a good idea of how many *coincidences* are going to be there as the result of chance alone—one can build up a picture of the "background." A true signal will show itself as an excess in the number of genuinely coincident pulses above the background estimates generated from the time slides.

A time slide can also be called a "delay." The signal will appear, in the language of Weber, as a "zero-delay" excess. Nowadays scientists look not for a zero-delay "excess" but at isolated coincidences between signals from different detectors. Nevertheless, the calculation of the likelihood that these coincidences could be real rather than some random concatenation of noise is based on an estimate of the background done in a way that is close to the method that Weber pioneered.

As the 1960s turned into the 1970s Weber published a number of papers claiming he had detected the waves, while other groups tried to repeat his observations without success. By about 1975 Weber's claims had largely lost their credibility and the field moved on. Weber's design of detector continued to be the basis of most of the newer experimental work, but the more advanced experiments increased the sensitivity and decreased the background noise in the "bars" by cooling them with liquid helium. Most of the experiments were cooled to between 2 degrees and 4 degrees of absolute zero, with one or two teams trying to cool to within a few millidegrees of absolute zero. Collectively, such "cryogenic bars" were to be the dominant technology in the field until the start of the 2000s. Just two groups, one based in Frascati and sometimes known as the "Rome Group" or "the Italians," and an Australian group, kept faith with Weber's claims, promulgating results that most gravitational wave scientists believed were false—the latter view being one which would now be almost impossible to overturn.

Nearly everyone outside the maverick supporters of Weber came to believe that Weber had either consciously or unconsciously manipulated his data in a post hoc way to make it appear that there were signals when really he was really seeing nothing but noise. This can happen easily unless great care is taken. Weber did not help his case when he made some terrible mistakes. In the early days he claimed to have a periodicity in the strength of his signals of twenty-four hours when proper consideration of the transparency of the Earth to gravitational waves suggested that the right period should have been twelve hours. Somehow, shortly after this was pointed out, the period mysteriously became twelve hours in Weber's discussions and papers, and this led some people to be concerned about the integrity of his analysis.[3] He also found a positive result that should have been ruled out because it was caused by a computer error, and, most damningly, he claimed to have found an excess of zero-delay coincident signals between his bar and that of another group when it turned out that a mistake about time standards meant that the signal streams being compared were actually about four hours apart, so that no coincidences should have been seen.

Those who had faith in Weber's experimental genius were ready to accept that these were the kind of mistakes that anyone could make, but

3. See chapter 5, note 6, below, for a retrospective reexamination of this criticism.

those who were less charitable used the events to destroy his credibility. Weber did his case further harm by the way he handled these stumbles. Instead of quickly and gracefully accepting the blame, he tended to try to turn it aside in ways that damaged his credibility. Weber's reputation fell very low, and the community tried to convince him that he should admit that he was wrong from start to finish, allowing them to give him more credit for his adventurous spirit and his many inventions and innovations, but he never gave in.

Weber died in the year 2000, insisting to the end that his results were valid and even publishing a confirmatory paper in 1996—a paper which nobody read. Weber was a colorful and determined character without whom there would almost certainly be no modern billion-dollar science of gravitational wave detection. I have heard Joe Weber described as hero, fool, and charlatan. I sense his reputation is growing again, as it has become easier to give credit to his pioneering efforts now that he is no longer around to argue with everyone who doesn't believe his initial findings. I believe he was a true scientific hero and that his heroism was partly expressed in his refusal to admit he was wrong; believing what he did, a surrender for the sake of short-term professional recognition would not have been an "authentic" scientific act. That the published results that indicate that he detected the waves are almost certain to remain in the waste bin of physics is another matter.[4]

For a time Weber was one of the world's most famous scientists, thought to have discovered gravitational waves with an experiment that was an astonishing *tour de force*. Many scientists now see the Weber claims as having brought shame to the physical sciences. Much of the subsequent history of gravitational wave detection has to be understood in the light of what happened.

Long after most scientists considered Weber to have been discredited, a group based near Rome, which will be referred to frequently in this book, published or promulgated several papers claiming to see the waves. The claims were based on coincidences between a cryogenic bar in Rome and one in Geneva, on coincidences with the cryogenic bar in Australia, and on coincidences between one of their original room-temperature bars

4. But there will doubtless be attempts to revive them. An entry submitted on 2 March 2009 on the arXiv blog, entitled "Were gravitational waves first detected in 1987?" (arxiv.org/abs/0903.0252), reports a paper by Asghar Qadir that has the potential to revive one of Weber's claims. I was to discover that this was another one of those papers which, though it has enormous potential to change things if true, was simply ignored by the gravitational wave physics community. For a related event in respect of one of Joe Weber's papers, see *Gravity's Shadow*, chapters 11 and 21.

and Weber's room-temperature bar.[5] These claims were sometimes ignored by the rest of the gravitational wave community and sometimes greeted with outrage.

The outrage, I believe, and have attempted to show in my more complete history of the field, can be to some extent correlated with the need to get funds to build a new and much more expensive generation of detectors. These are the interferometers which today dominate the field. An experiment like Weber's could be built for a hundred thousand dollars, whereas the U.S. Laser Interferometer Gravitational-Wave Observatory (LIGO) started out at around a couple of hundred million. If gravitational waves could be detected for a fraction of the price, and Weber once wrote to his Congressional representative to argue just this, funding for big devices would be hard to justify. Therefore, it became a political as well as a scientific necessity to stress that the bars could not do the job that Weber and the Rome Group were claiming for them. On the basis of almost every theory of how these instruments worked, the interferometers were going to be orders of magnitude more sensitive than the bar detectors, and on the basis of almost every theory of the distribution and strength of gravitational wave sources in the heavens, only the interferometers had any chance of seeing the waves. Furthermore, even the first generation of these more expensive devices could not be expected to see more than one or two events at best. The consensual view among astrophysicists was that the sky was black when it came to gravitational radiation of a strength that could be seen by the bars, including the cryogenic bars, and that it might emit a faint twinkle, perhaps once year, as far as the first generation of interferometers was concerned. The promised age of gravitational wave astronomy, involving observation of many different sources with different strengths and waveforms, helping to increase astrophysical understanding, would not be here until a second or third generation of interferometers were on the air. It was only the promise of gravitational astronomy, not first discovery, which could justify the huge cost of the interferometers.

Thus the scene was set for the unfolding of a battle between the cryogenic bars and the interferometers, with the bar side led by the Rome Group. Some bar teams, such as those based in Louisiana and in Legnaro, near Padua, accepted the view of the interferometer teams and agreed to strict data analysis protocols based on a model of the sky in which signals would be rare and strong. This ruled out any chance of detecting weak

5. Weber never constructed a successful cryogenic bar, though he was given funds to embark upon the project.

signals near the noise that might otherwise have been used as a basis for tuning the detection protocols. This was the bars' last chance—it was probably the only way they could work toward an understanding of any weak signals good enough to survive the more severe statistical tests needed for a claim.[6] But the Rome Group was not prepared to accept the dismal astrophysical forecasts and exerted the experimentalist's right to look at the world without theoretical prejudice. If the Rome Group could work themselves into a position that enabled them to find some coincidences that could not be instantly accounted for by noise, and which flew in the face of theory, then they were determined to say so—in the spirit of Joe Weber. They were not willing to put all their effort into explaining away every putative signal just because it was supposed to be theoretically impossible. Thus were they to give rise to a continuing history of "failed" detection claims right into the twenty-first century, and thus did they give teeth and muscle to the history-monster stalking Arcadia's corridors.

Interferometers

Five working interferometers play a part in this story. The size of an interferometer is measured by the length of its arms. The smallest, with arms 600 m long, is the German-British GEO 600, located near Hannover, Germany. Virgo, a 3 km French-Italian device, is located near Pisa, in Tuscany. The largest are the two 4 km LIGO interferometers, known as L1 and H1, located respectively in Livingston, Louisiana, close to Baton Rouge, and on the Hanford Nuclear Reservation in Washington State. There is also a 2 km LIGO device, H2, located in the same housing as H1.

An interferometer has two arms at right angles. Beams of laser light are fired down the arms and bounced back by mirrors. The beams may bounce backward and forward a hundred times or so before the light in the two arms is recombined at the center station. If everything works out just so, the changing appearance of the recombined beam indicates changes of the lengths in the arms relative to each other — a change that could be caused by a passing gravitational wave. It should thus be possible to see the "waveform," of a passing gravitational wave in the changing pattern of light that results from the recombination of the beams.

6. This is roughly what happened in the case of the 2002 publication by Astone et al. (see below). In that case, however, the tuning led to their *not* finding a signal in a second period of observation, thus experimentally *disproving* their tentative claim—which seems a reasonable way to do science.

The longer the arms are, the larger the changes in arm length and the easier it is to see them, so, other things being equal, bigger interferometers are more sensitive than smaller ones. But even in the largest interferometers, the changes in arm length that have to be seen to detect the theoretically predicted waves would be around one-thousandth of the diameter of an atomic nucleus (i.e., 10^{-18} m) in a distance of 4 km. It is, therefore, something close to a miracle that they work at all, where "working" does not necessarily mean detecting gravitational waves but being able to measure these tiny changes.

LIGO was funded in the face of bitter opposition, some from scientists who believed the devices could never be made to function. I was lucky enough to watch every stage of the building of the LIGO interferometers, and for much of that time I too did not believe they were going to work, and I was not alone even among those close to the technology. To see the first tentative indications that the trick might be pulled off, and to watch the slow increase in sensitivity right up to design specification, two or three years late though it was, has been one of the most exciting experiences of my life—perhaps more exciting that even the final detection of gravitational waves will be. But even now the big interferometers are far from perfect machines—they are still plagued by undiagnosed sources of noise which, as we will see, make their effective range somewhat less than the range as calculated from the moment-to-moment performance of their components.

Range is vitally important. Astrophysical events that might be visible on an Earth-bound gravitational wave detector are unpredictable and may happen anywhere that galaxies are found. The greater the range, the more galaxies can be included in the search and the better the chance that a wave will be seen. The number of galaxies, and therefore the number of potentially exploding or colliding stars that might be seen, is proportional to the *volume* of space that can be surveyed. This volume is a sphere centered on the Earth, and the number of stars and galaxies it contains is roughly proportional to the cube of the radius—the radius being the range. Thus, a small increase in range buys a proportionally much greater increase in potential detections; if the range is doubled, the number of potential events increases by eight; if the range is multiplied by ten, as is the promise for the next generation of LIGO detectors, the number of potential sources will be increased a thousand-fold. When this happens the promise of gravitational wave *astronomy* might be fulfilled.

GEO 600, because of its relatively short arms and some other problems, does not play much part in the story to be told here. Virgo, even though its arms are only 3 km long rather than the 4 km of L1 and H1, the big LIGO devices, includes clever aspects of design that should give it a better potential performance at low frequencies; this makes it a more important contributor to the detection process than its higher frequency performance alone would imply. Unfortunately, progress on the low frequency aspect of the design was slow—low frequency is always more difficult—and, in general, its development has suffered greater delays than expected and so its sensitivity has lagged further behind that of LIGO than technical limitations would imply. This will turn out to be indirectly important to the story.

LIGO's 2 km interferometer, H2, is an anomaly. A crucial element of a detection claim is the existence of coincident signals between widely separated detectors. GEO 600 is near Hannover, Virgo near Pisa, L1 is close to Baton Rouge, and H1 is on the Hanford Nuclear Reservation in Washington State. But H2 is located in the same housing as H1, so there is no separation involved. This makes coincidences between H2 and H1 much harder to interpret as a signal than coincidences between any of the other pairs of detectors. H2 nevertheless plays a part in the story.

LIGO is now known as "Initial LIGO," or "iLIGO," because one-and-a-half further generations are in progress. The half generation is Enhanced LIGO (eLIGO)—LIGO with certain components of Advanced LIGO (AdLIGO) installed early. If it does all that is intended of it, eLIGO will have twice the range of LIGO and be able to see eight times as many potential sources. eLIGO is just coming on air, though at the time of writing it has some troubles that are pushing back the start of double sensitivity and may even put its achievement in doubt. AdLIGO is to be installed in the same vacuum housings as Initial LIGO but has all new components, including better mirrors, better mirror suspensions, better seismic isolation, and a much more powerful laser. AdLIGO should be producing good data around 2015. Already I have heard people saying that eLIGO is justified by its role as a test bed for AdLIGO components or even that this is all it was intended to be in the first place. My recollection of the arguments for its construction is that the strongest pressure came from scientists who believed doubling the sensitivity would be the key to the initial detection of gravitational waves, as Initial LIGO, for which some had held out great hope, was proving a disappointment. In the absence of pressure from some senior scientists who were sure that eLIGO would produce the desired result, it seems possible that the device would not have been built and that the initial run of LIGO would have been extended instead of disassembling the machine so soon for the new components to be installed. On the other hand, I think it is also the case that eLIGO would not have gone ahead if it had needed components that were not to be installed anyway for the use of AdLIGO, so that it could be used as an AdLIGO test-bed as well as a detector in its own right.[7] I get the sense that the scientists will feel that honor has been satisfied, or perhaps more than satisfied, if the total number of potential sources surveyed by eLIGO equals or exceeds the number that would have been observed if Initial LIGO had stayed on air until all the components were ready to be installed in AdLIGO. The calculation of how many sources have been observed involves integrating the time of observation multiplied by the cube of range. In other words, if eLIGO does achieve twice the range of LIGO, then one month of eLIGO's time is worth eight months of LIGO's (but should be worth only about six hours of AdLIGO's).

A feature of these calculations is the duty cycle. An interferometer is not always in a state to make observations when it is "switched on." First, there are periods of necessary maintenance. Then there are periods when the detector cannot achieve "science mode" because the environment is too noisy. Noise can come from what the scientists like to call anthropogenic sources—airplanes fly low over the site, piles are driven or pneumatic drills are used, trucks come close when delivering supplies, in Louisiana, trains pass, logs are felled, and, disastrously, explosive devices are fired for oil or gas prospecting—an eventuality which is going to require the shutting down of the entire detector for a month or two toward the beginning eLIGO's run. Downtime can also be caused by natural events. Seismic disturbances big enough to shake the detectors out of science mode can be caused by earthquakes, while lesser noises are caused by storms which pound the shores with big waves or winds which shake buildings or drag on the ground.

There are a number of levels of disturbance that affect an interferometer. The most severe of these is when the device goes "out of lock." The mirrors in an interferometer must be isolated from the crude shaking of the ground which, if they were fixed in place, would disturb them trillions of times more than the effect of a gravitational wave. The mirrors are therefore slung in an exquisite cradle of pendulums and soft vertical isolation springs all surrounded by hydraulic feedback isolators to cancel out large external disturbances. "Lock" is a state where the necessary spider-

7. Jay Marx pointed out to me (private communication October 2009) that the extra budget allocated specifically for the putting together of eLIGO was only $1.4 million, so, in the context of the whole program, it seemed the right thing to do given the extra chance of making a detection.

web of feedback circuits which control the oscillation of the mirrors are in a degree of balance such that that mirrors are stationary and the laser light can bounce between them, building its strength, and be used to measure the relative lengths of the two arms with maximum accuracy. When the interferometer is in that condition, any extra impulses the interferometer's electronics has to send to the mirrors to hold them still is a measure of the signal that is being looked for. In other words, if the mirrors are perfectly balanced in a state which is undisturbed by extraneous forces, any tendency to move *may* be caused by the caress of a gravitational wave, and the tiny restoring forces required to prevent the mirrors being affected by this caress is a measure of the wave's form and strength. When the interferometer is badly disturbed, the shaking of the mirrors becomes so great that the feedback circuits cannot deliver enough power to the actuators to hold them in the position needed for the laser light to bounce between them; this is "going out of lock."[8] But even when the device is "in lock" there are periods when the data is useless or compromised. Once more, the major source of this problem is outside disturbance: even if the mirrors can maintain lock, there may be so much going on in the feedback circuits that the effect of any gravitational wave would be swamped. To sense the caress of a gravitational wave the interferometer must be asleep, not tossing and turning. An interferometer, like all gravitational wave detectors, is like the princess in "The Princess and the Pea," while a gravity wave is the pea. The pea will only disturb the princess's sleep in a meaningful way if that sleep would otherwise be deep and peaceful.

Unwanted lumps in the princess's mattress, that is, unwanted disturbances to the interferometer, are recognized by the network of "environmental monitors" checking seismic, electrical, acoustic, and every other imaginable kind of intrusion, including those caused by internal problems such as jitters in the laser, stray light affecting the feedback circuit sensors, or any departure from stability in the vacuum system. When any of these problems arise, a "data quality flag" is posted; stretches of data stigmatized by flags will, in most cases, be "vetoed." Once more, in an ideal world, the entire process would be automated: the monitors would note trouble, write the flags, and those stretches of data would be thrown away. But, again, human judgment cannot be dispensed with. How serious is a veto? Too low a bar for the posting of a flag and so much of the data would be

8. Virgo has a much better duty cycle than the LIGO detectors. This may have something to do with its clever and complex mirror suspensions which might make it less subject to the effects of external disturbances which can cause an interferometer to go out of lock.

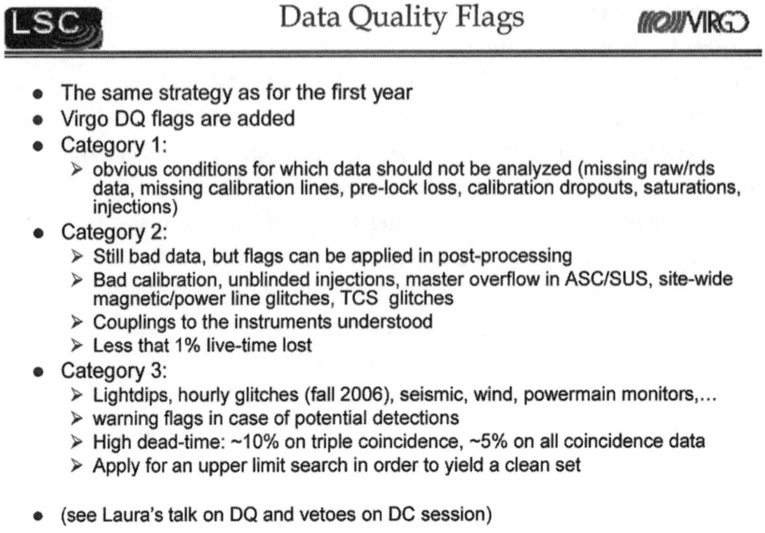

Data Quality Flags

- The same strategy as for the first year
- Virgo DQ flags are added
- Category 1:
  - ➢ obvious conditions for which data should not be analyzed (missing raw/rds data, missing calibration lines, pre-lock loss, calibration dropouts, saturations, injections)
- Category 2:
  - ➢ Still bad data, but flags can be applied in post-processing
  - ➢ Bad calibration, unblinded injections, master overflow in ASC/SUS, site-wide magnetic/power line glitches, TCS glitches
  - ➢ Couplings to the instruments understood
  - ➢ Less that 1% live-time lost
- Category 3:
  - ➢ Lightdips, hourly glitches (fall 2006), seismic, wind, powermain monitors,...
  - ➢ warning flags in case of potential detections
  - ➢ High dead-time: ~10% on triple coincidence, ~5% on all coincidence data
  - ➢ Apply for an upper limit search in order to yield a clean set

- (see Laura's talk on DQ and vetoes on DC session)

Figure 1. Three levels of veto

thrown out that the duty cycle would be reduced and reduced and that first discovery might be lost among the discards. Too high a bar and the results will be unreliable. The fact that judgment is central is institutionalized in the different way vetoes are handled for upper limits and discovery candidates. For upper limit claims, vetoes are less strict than for discovery candidates. These are the conservative courses of action. Furthermore, three categories of veto of different degrees of seriousness are set up. Figure 1 is a PowerPoint slide from a meeting which summarizes the characteristics of the three types of veto.

Of course, judgment was used to decide on the categorization—the machine could not make categories for itself. With those judgments made, it is expected that data affected by the least serious "category 3" vetoes, though it will not figure in the first analysis, will be examined later if other eventualities indicate that an interesting signal might be in danger of being ignored because of what is meant to be a "light touch" warning. This element of judgment will figure in the story to follow.

Taking all this into account, Initial LIGO had to run for about two years to gather one consolidated year's worth of data. There had been four prototype "science runs" at less than design sensitivity, so this final two-for-one-year run was known as "Science Run 5" or "S5." Enhanced LIGO's science run will be known as "S6," though there are intimations at the time of writing that there will be an S6a at well below design sensitivity and an S6b that will be nearer to what had been hoped for. Thus, when we say that if S6 reaches its design sensitivity, one month of running will be worth eight months of S5, where the "months" are integrated periods of operation in science mode. The abbreviations, H1, L1, H2, S5 and S6, and their meanings, should be remembered. They will be used repeatedly.

The range of an interferometer, at least for H1, L1, and H2, the three American interferometers, is expressed in a standardized form. A typical source of gravitational waves would be the final moments in the life of a pair of binary neutron stars. The orbits of binary stars slowly decay, and as they get closer and closer to each other they circle faster and faster (like skaters drawing in their arms). In the last few seconds before the stars merge they will circle each other hundreds, then thousands, of times a second in a rapidly increasing crescendo which, if you could hear it, would sound like a "chirp." The standardized range is the distance at which such a system, with both components being 1.4 times the mass of our Sun, and orientated in neither a specially advantageous nor a specially disadvantageous way, ought to be seen by the detector. Since whether it would be seen or not depends on the momentary state of lock, noise, and data quality, the range of any detector is continually fluctuating. From the middle to the end of S5, L1 and H1 had ranges of around 14 to 15 megaparsecs (a

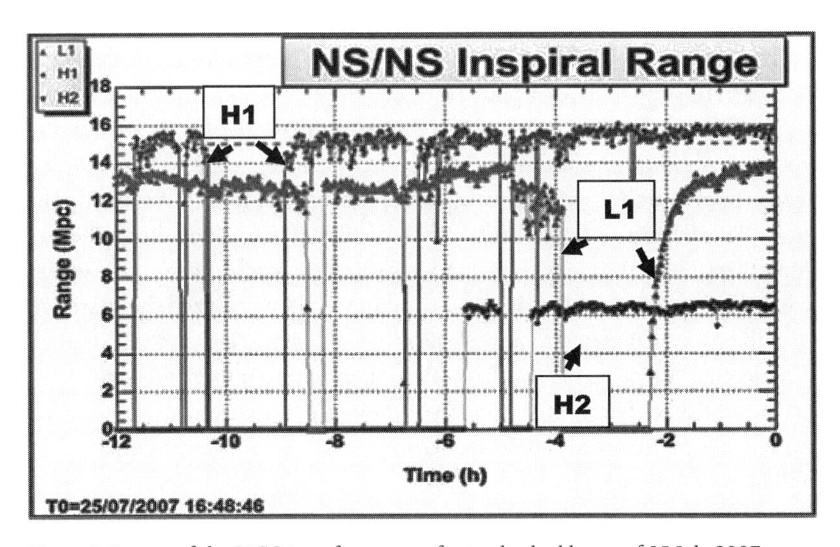

Figure 2. Ranges of the LIGO interferometers for twelve bad hours of 25 July 2007

megaparsec is a little over three million light-years) with H2's range being about half of this—as would be expected given that it is half the length of the other two. This state of affairs is represented in figure 2, which is taken from the electronic logs maintained by the laboratories.

On most days the three plots would run nearly all of the way across the graph, and the two upper ones would be almost superimposed at the 14–15 Mpc range. It is easier to disentangle the two plots on this bad day because one of the detectors has a range of only around 13 Mpc. H2, as can be seen, is not having a great day either, being down for more than half the time and having a range of only just over 6 Mpc for the rest of the day, whereas something around 7 would be more typical.

Measured by the standard of LIGO on a good day, eLIGO was, therefore, intended to have a range of about 30 megaparsecs. At the time of writing I hear scientists expressing the hope that it will reach 20 before it has to be switched off for the installation of AdLIGO components.

Data Analysis

What this introduction to gravitational wave detection science should have made clear is that taking data from a gravitational wave detector is not like reading the electricity meter. It should be obvious that whether some slightly unusual activity in the interferometers has the potential to count as the result of the touch of a gravitational wave depends upon whether it is coincidental with activity on at least one other detector. If there was a third, fourth, or fifth detector with enough sensitivity for the event to be within range, then either the event must be seen in them too or there must be a good explanation for why it was missed. The explanation will depend on decisions about the scientific state of the detectors involved—whether they were in a sufficiently peaceful state to be disturbed by the pea, if such it was. As we will see, these considerations are only the start. There are subtleties upon subtleties involved in deciding whether something is sound data. These considerations reach out all the way to what the scientists were doing and thinking as they analyzed the material. As will be explained in chapter 5, this is because statistical analysis, though it looks like the purest of mathematics when on the published page, turns on unpublished details of the analysts' thoughts, activities, and statements from decades previously, and reaches forward to what they anticipate they will be doing decades hence.

A detection, then, however it is dressed up, is not a self-contained "reading." It is not Nature, reflected clean and pure on the still surface of science's pool; there is no pool, only the fast flowing river of human activity on a restless Earth. The first detection will be a social and historical eddy—at best a momentary stillness in the white-water of history.

The story of this book turns on a planned rehearsal of stillness-making. It concerns a deliberate test of the LIGO team's abilities, staged toward the end of S5, and known as the "blind injection challenge." Two scientists were charged with injecting zero, one, two, or three false signals into the LIGO interferometers with unknown form and magnitude according to a sequence of random numbers: the job of the team was to find them if they were there, always being aware that there might be none at all, so that any putative signal might be real.[9] The story of that adventure is the spine of the book.

9. Allan Franklin (private communication, October 2009, referring to his 2004) points out that blind injections are by no means unprecedented: "Consider the episode in which it was claimed that a 17-keV neutrino existed. One of the reasons why the Argonne experiment (Mortara, Ahmad et al. 1993) was so persuasive was that they were able to demonstrate that they would have detected such a heavy neutrino had it been present as a result of their proven success in detecting similar blindly injected events." One of the purposes of the gravitational wave blind injections could be said to "calibrate" the whole detection procedure.

2 The Equinox Event: Early Days

For forty years and more, gravitational wave detection physics has been full of excitement. There were the early claims of Weber, there were the disputes surrounding "the Italians," and there was the near-miracle of the big interferometers being built and made to work. Strangely—and I found some of the scientists themselves were of the same opinion—it has been a bit dull since the 2002 Rome Group's claims were crushed and the interferometer data started coming in. Now that the machines are doing what they are built to do, things have become routine. What has been going on is a huge and widely distributed effort of data analysis, none of it showing any sign of a gravitational wave. The data hasn't gone to waste, as it has been used to set a sequence of upper limits: a whole string of papers has been published which show that the maximum flux of gravitational radiation of this sort or that sort is, as expected, less than "X." As we will see, upper limits keep the science going, and occasionally they have real astrophysical significance, but to an outsider like me, and to a proportion of the insiders, they are pretty boring. *"As scientists predicted, LIGO fails to see anything,"* does not make much of newspaper headline.

Suddenly, in the autumn of 2007, at the time of the September equinox, for outsiders like me, and for a lot of the insiders, life in gravitational waves became exciting again. The "Equinox Event" is a coincident pulse of energy the LIGO interferometers detected on 21/22 September 2007. It rapidly became clear that it might constitute the first discovery of a gravitational wave.

The Data Analysis Groups

At this point we need to turn aside from the main thrust of this chapter and explain a little more about how the data analysis is organized. This is necessary if the tensions and responsibilities surrounding the Equinox Event are to be understood.

Interferometric gravitational wave detection has the potential to spot four kinds of signal. Aside from their absolute sensitivity, the biggest advantage of the interferometers over the bar detectors is that they are "broad band.'" The bars could do little more than sense when a gravitational wave "kick" impacted on them; they would sense this as a net increase in the energy contained within the vibrations of the metal. But an interferometer should be sensitive to the shape of the signal as well as the aggregate energy it delivers. The movement of the mirrors should follow the actual pattern of the wave as it causes its distortions in space-time. Thus, for example, if the signal is caused by the final seconds of the inspiraling of a binary star system, then the very details of the "chirp" caused by the rapidly increasing rotation of the two stars around each other will be inscribed on space-time and in turn expressed in the rapidly changing separation of the mirrors (or, to be technically more exact, the rapidly changing feedback signal needed to hold the mirrors still in the face of the forces they experience). Such a signal should, then, be easy to identify because the same clear waveform should be expressed, more or less simultaneously, on the two or more widely separated detectors.

As always, real life is much more complicated because the new wave will be superimposed on the ever-present noise in the detectors and will need to be extracted by sophisticated algorithms instantiated in computer programs. The most important part of the technique is "template matching." A huge bank of many thousands of signal templates, corresponding to different inspiral scenarios, has been constructed. Thus one template will refer to the expected pattern of two 1.4 solar mass neutron stars spiraling into each other; another template will refer to a 1.4 solar mass neutron star and a 10 solar mass black hole spiraling into each other; another will be for two 20 solar mass black holes, and so on for as many combinations as one has computer power to manage. Since any specific signal might well fall between two templates and be mixed up with the noise of the detector, one can see that the technique is not as neat and tidy as it appears when first described. The duty of analyzing such signals has been given to a dedicated group known as the "Compact Binary Coalescence" or "CBC" group, which is more widely known by its original title: the "Inspiral Group." We'll call it by this name from here on.

Any asymmetric spinning star will also emit gravitational waves. Therefore another kind of signal that has the potential to be seen is that produced by pulsars—rapidly spinning stars that emit a beam of energy (which can be seen in its regular sweeps across the Earth) and that, therefore, can be assumed to be asymmetrical to some extent. The frequency of the gravitational wave signal produced by a pulsar will be the double the frequency of the flashes of light caused by the sweep of its beam. Therefore, if there is enough asymmetry to produce a large enough flux of gravitational waves, it should appear as a regular pulse in the interferometer. The "Continuous Wave" group has the job of looking for this kind of signal.

A third kind of signal is that left over from the initial formation of the Universe. It is the gravitational wave equivalent of the famous, electromagnetic, cosmic background radiation. Astrophysicists and cosmologists are particularly excited about this kind of signal because while the electromagnetic background goes back only so far toward the Big Bang, the gravitational wave background goes back to almost the very beginning. The form of this signal is random rather than following the pattern of a definable chirp or the regular heartbeat of continuous waves. It is known therefore as the "stochastic background." It can be detected by long-term correlations in what looks like "noise" in the detectors but is really the signature of the random background waves.

The fourth kind of signal is a burst of unknown origin or form. It might be emitted by a supernova, by an inspiraling system with an awkward signature, or by something not yet understood. Searching for these signals is the job of the "Burst Group."

First Intimations of the Equinox Event

It was the Burst Group which found the Equinox Event. Over the years I have slowly come to gain more and more access to the goings on the gravitational wave detection community, and nowadays I am allowed to listen in to the weekly telephone conferences, or "telecons," of the groups. Mostly these are highly technical and I listen to only a few of them. The first I heard of the Equinox Event was when a member of the Burst Group emailed me on 2 October suggesting that I pay special attention to the telecon coming up on the 3rd. He said:

> This should be an interesting telecon for you to sit in on, because there was a notable event candidate found by one of our online searches in the data from Sept. 21, and brought to the attention of the working group during last week's call. We will be talking more about it and planning to start working through our "detection checklist."

This was exciting.

Nothing is simple, and the first thing to understand is what is meant by "one of our online searches." The gravitational wave analysts are especially sensitive to the possibility that they might be accused of biasing their data by post hoc choice of search parameters, or "tuning." Such sensitivity is good practice in all sciences, many of which try to avoid it by "blind" analysis, but, as we have seen, the history of the gravitational wave field has made it an issue with a very high profile.

Unfortunately, however, data analysis is just like experiment in that a fair amount of trial and error is involved. The idealized model would hold that the exact recipe for data analysis could be worked out in advance before anything is actually analyzed, but the truth of the matter is that data analysis is long and subtle and things always go wrong. The problem is nicely illustrated by the subject line of an email that was circulating in April 2009: "*Re: [CBC] 12TO18 Re-re-re-re-re-re-re-re-rerunning our upperlimits because of this V4 calibration issue.*" Furthermore, much can be optimized only after a series of trial runs. The solution that the gravity wave teams adopt to resolve this problem is to tune the procedure on data that is not going to be used for the main analysis. One approach is to split the data into two sections, a small one—around 10 percent of the whole selected at random—and the main section. The small section is known as the "playground," and anyone is allowed to do anything they like with it, including any amount of post hoc readjustment of parameters, until they work out the best algorithm they can for extracting potential signals from noise. Another approach is to do the tuning on data where the output from one detector has been subject to a time slide, so that anything to which the procedure is tuned cannot be a real signal. Only when all this work has been completed in one way or another, and the data analysis protocol is "frozen," is the "box opened" on the main body of data or on the data which has not been subject to a time slide. The main body is then analyzed with the frozen protocol, which must not be changed in any way. We might call this procedure a "meta-rule" of data analysis. The meta-rule is that nothing can be touched or adjusted after "playtime" is over; once the box is opened the analysis rules stay as they are.

Opening the Box

Jumping ahead for the sake of exposition, I was present when the box was opened by the Burst Group on the first calendar year of their data. This seems a good place to describe what it is like to "open a box," though this description is an indulgence, playing no part on the main theme of this chapter or of the book.

At 8 a.m. on 16 March 2008 about fifty people sat scattered in a dimly lit lecture theater on the campus of the California Institute of Technology. Everyone, as is nowadays customary at such gatherings, was alternately listening to the speaker and staring at, and perhaps typing into, the glowing notebook computers on their laps. Each notebook was linked by WiFi to the Internet. Peering over scientists' shoulders, one might see email being dealt with, papers scanned or composed, or lines of computer code written or debugged. The discussion centered on whether the protocol had been finalized and whether there was anything more to do before the moment came from which they could not turn back. Whatever was not done before the box was opened could not be rectified afterwards. There ought to be great drama. Would the refined algorithms set loose on the first significant body of data from "the age of interferometry" in gravitational wave detection physics see anything?

The moment when it came was strangely downbeat. Everyone agreed that nothing had been forgotten, and the scientists sitting in the rows of seats were given permission to open the box. The box was opened by some of them pressing a few buttons on a portable computer—looking around you could not see who was doing the work and who was not. The message was carried at the speed of light back to the banks of computers located at the main sites. A few moments later the entire database was analyzed, and the result flashed back via the satellites servicing the Internet. The scattered scientists with the portable computers reported that nothing of significance was found. An analysis that required hundreds of millions of dollars, subtle political skill, and virtuous and sometimes brutal management to enable, decades to plan, years of data collection, and months and months of agonizing over whether all the preliminary tasks have been properly done, was completed almost as soon as it was started and there was nothing to show. This is physics.

The "Airplane Event"

How seriously the distinction between playground and main data is taken is illustrated by a bizarre data analysis incident now referred to by everyone in the collaboration as "the airplane event." One only has to say "airplane event scenario" to a member of the collaboration and they immediately know what is at stake.

In mid-2004 the Burst Group were preparing an upper limits paper. Upper limits are a kind of scientific finding that was described in *Gravity's Shadow* as turning lead into gold. When gravitational wave scientists fail to see any gravitational waves, they turn this into a finding by saying it proves that the upper limit on the possible flux of the waves is "such and such." Since LIGO first went on air, a stream of these upper limits papers have been published. One or two of these have been of some astrophysical interest. For example, there have been papers putting upper limits of the gravitational wave component of the energy lost in the slowing down of the rate of rotation of the Crab pulsar. Assuming all the assumptions on which gravitational wave science is based are correct, these set bounds on the degree of asymmetry of this particular neutron star. Again, when a powerful gamma ray burst was seen in the direction of the Andromeda Galaxy, a paper was promulgated on ArXiv, the electronic preprint server, in November 2007, with the following included in the abstract:

> We analyzed the available LIGO data coincident with GRB 070201, a short duration hard spectrum gamma-ray burst whose electromagnetically determined sky position is coincident with the spiral arms of the Andromeda galaxy (M31). . . . No plausible gravitational wave candidates were found within a 180 s long window around the time of GRB 070201.

This was taken to show that if the event had been caused by an inspiraling neutron star/neutron star system, or an inspiraling neutron star/black hole system below a certain size, it could not have been located in the Andromeda Galaxy but must have been well behind it even though it was in the same line of sight.

Such results are marginally interesting, and by all accounts, were well received at the astronomy conferences where they were presented, but most of the upper limit papers seem rather boring now that the excitement of being able to say anything at all about gravitational wave fluxes had worn off. Most of these papers say something like "you expected there to be less than flux X of gravitational waves of this kind: we've proved there's less than 100X." The "100X" gets steadily lower and lower as the instruments become more sensitive, but, mostly, they are still a long way from saying anything that is going to cause astrophysicists to start scratching their heads.

In the world of gravitational wave detection, the assiduously honest analyst will be leaning over backwards to make sure that nothing that is not a gravitational wave is announced as being a gravitational wave. In the mirror-image world of upper limits, the same analyst will lean over backward to make sure that nothing that could be a gravitational wave is said *not to be* a gravitational wave. This is because excluding too much gives a too low (and, therefore, a misleadingly interesting!) upper limit. It is just as important in setting upper limits that the meta-rule is followed and that no potential gravitational waves are eliminated by jigging around with the analysis procedure once the box is opened. It is a violation of the meta-rule in the case of the upper limit paper of 2004 that gave rise to the bizarre incident.

The violation was the following: after the box had been opened, a big event was found, but then someone noticed that the loudest event was actually correlated with an airplane whose passage over the site had been recorded on a microphone. Unfortunately, the possibility of low-flying airplanes giving rise to false signals in the data stream had not been envisaged during the playtime period, so the protocol did not allow for such things to be removed. It was clear that this event could not be a candidate to be a gravitational wave, but to remove it post hoc, "by hand," as it were, would violate the meta-rule for setting upper limits: it would *improve* the upper limit as a result of a post hoc procedure. But to leave it in would mean that the upper limit paper contained an artifact that made it incorrect. The debate was about whether the meta-rule should be violated or a false result published. This debate had been going on for about four months when, in November 2004, I witnessed one of its final paroxysms. The following are anonymized extracts from a heated discussion among about sixty people that lasted approximately half-an-hour and which, because it could not be resolved, ended in a vote (!) on what should be done.

S1: We've got a positive not-gravitational wave event and I don't understand why it's still in the analysis.

S2: It is possible to remove it but it is a moral and philosophical question.

S3: In August when we discussed it, the data analysis group had a show of hands. The risk in this case is that you make the mistake that the playground is designed to keep you from making—which is that, retrospectively, you take the airplane to have caused the event but it could still be a gravity wave coincident with the airplane. Up to a week ago I was not sure but in fact I believe now that the airplane really did cause the event. [Shows plots of trace from microphone and trace from detector.] You can see that the loudest time in the microphone does not correspond to the event but clearly this stretch of data here corresponds to this stretch of data here and so its pretty clear that this stuff is being caused by the airplane, and if you look at this loud noise here—if you look at the frequency content of that—you find that it is around 85 Hertz—it is around the frequency of the airplane sound at that time. . . . I'm willing to state that I'm positive that the event was actually caused by the airplane and this event can be tied to that physical effect and that I would personally be more comfortable now than I was after the initial study about throwing that out.

S4: I am in favor of leaving the event in the upper limit calculation, and stating, of course, clearly in the paper that this is not a real event—that it has an environmental cause—but I am against changing the upper limits that we're quoting and it's because I don't know quite enough about statistics and bias to convince myself that this is a safe procedure—to go and change the algorithm by which we compute the upper limit after . . .

S5: It was always anticipated that one would check environmental signals!

S4: Yes, for vetoing a detection claim, but we did not discuss and we haven't really thought deeply about using that to change an upper limit.

S6: We did discuss it long and ago—we decided we weren't going to do that.

S4: . . . the upper limit that we quote is only meaningful when you have statistical confidence limit associated with it, and if we introduce a bias that leaves that confidence in doubt then your upper limit is meaningless. . . . We have to be conservative and take the hit on the upper limit rather than risk quoting an upper limit that's actually incorrect.

S7: If we have confidence beyond a reasonable doubt that we have an airplane, then I think we should interpret our results in that context. If someone had thought about it beforehand and just not mentioned it and then steps forward we would have no hesitation about applying it now. We would not worry about any statistical bias that would result; otherwise it is too much mathematics and not enough physical reality.

S4: The flaw is when the person steps forward and said "I knew about this before." Because that changes the algorithm. If there hadn't been an airplane that person wouldn't have stepped forward, the change wouldn't have been implemented. I really feel very uncomfortable about changing these algorithms after we've looked at the data. I think the conservative thing is to just not touch it.

S5: I think there's a difference between a posteriori adjustments like this when you're making a claim for a detection and when you're making a claim for an upper limit. I think in this case, we're not making a detection claim—we're simply saying that there's something that we believe should be excluded from the gravitational wave search because it's contaminated in a very clear way by environmental influences, and it shouldn't be part of the upper limit. . . . I think we would look pretty ridiculous if we left it in. . . . Otherwise it's not the best we can do on the basis of the data. We're being paid to do the best job we can. I mean otherwise why spend all this money on these detectors if we're not actually going to do the best job we can with the data we have. . . .

S9: The truth is that the most conservative thing and the thing you know won't violate your statistical bounds is to leave the event in. You know that. But also the truth is that if, as a collaboration, we firmly believe that it originates with the airplane, then the bias it introduces by applying that veto over the entire analysis is minimal. The trouble is that if you did find a loud event like this and everyone didn't go back to the [plots of the traces] but you just throw it out, then you place too low an upper limit. But from what I can tell from listening to people who know the event and have studied it carefully, we should take it out because it introduces a small bias but is not going to put us into that scary regime of making too tight an upper limit.

S6: I'm convinced that this is an airplane, and if we had decided, six months ago, that we would hand-scan the events I would have no problem. The problem is that six months ago we agreed to not do that. If we remove it we're telling the gravitational wave community that we allow ourselves the right to change our analysis after the event.

S10: We may learn from our mistakes unless we agreed earlier not to, and in this case we agreed earlier not to learn from our mistakes!

S6: We agreed to learn from our mistakes, but not to attempt to correct them because we can correct them for the next round of data analysis without doing anything retrospective.

S3: Our draft paper currently says "We observed one event with an upper limit of 0.43 per day. By the way, this event was an airplane." It's not our best job of saying what is the limit on the gravitational wave rate.

S11: [If both limits were in] that would give the reader an opportunity to decide for himself.

Collins *(sotto voce)*: That's postmodernism.

S12: We're sure this is an airplane and we're sure the result we've produced is wrong.

S4: No—we're only sure the result we've produced is right if we leave the airplane there.

S6: It's a conservative upper limit.
S7: Let's not misuse the word "conservative" . . . It's mad!

Not long after this the meeting concluded that it would be impossible to reach a consensus and that they must vote on it. (There were many humorous remarks referring to the recent American Presidential election.) The vote went eighteen for leaving the airplane in and a much larger number (around thirty) for taking it out. The leave-it-in group was asked to concede, which they did, but not without at least some people feeling frustrated and certain that a poor decision had been made and not without one scientist insisting his name be removed from the published paper.

While everyone felt uncomfortable about taking a vote, to the extent of people looking round at me and laughingly acknowledging that "I had got what I came for," it was an embarrassingly long time later before I realized what it all meant. The procedures of science are meant to be universally compelling to all. To find that one needs to vote on an issue institutionalizes the idea that there can be legitimate disagreement between rival parties. In other words, taking a vote shows that there can be a *sociology* of science that is not rendered otiose by the universally compelling logic of science. Though we did not realize it at the time, a vote at a scientific meeting is a vote for sociology of science and we all felt it instinctively. It is not that there aren't disputes in science all the time, but putting their resolution to a vote legitimates the idea that they are irresolvable by "scientific" means; neither the force of induction from evidence or deduction from principles can bring everyone to agree.

There was one fascinating, *Alice in Wonderland*, moment during the debate when someone pointed out that, if the airplane was removed, the upper limit might be low enough to be said to conflict with the "Italian" 2002 claim—it would drive the upper limit down to the point where it ruled out the flux that the Rome Group had claimed to see. In response, a number of others opined that in that case it really would be *illegitimate* to remove the airplane because its removal could be said to be an interested after-the-fact maneuver. In that case the analysts would feel themselves very vulnerable to accusations of statistical bad faith. In other words, if the removing of this spurious piece of data had any real astrophysical significance, then it could not be removed because then it really would look like post hoc data massage; if it had no significant astrophysical significance then it could be removed!

Another fascinating feature of the debate was that two of the speakers who were most adamant in wanting the airplane removed, exposing the group to a potential statistical massage accusation, had each published papers complaining about the Rome Group's post hoc misuse of statistics. Ironically, other scientists in the room *took comfort* from the fact that even these persons, whose track record of adamant rejection of post hoc analysis had proved their statistical propriety, thought that the airplane should be removed. Strangely, no-one seemed to notice the irony.

It seems to me that the right way to proceed in this case was to remove the airplane event, since any other course of action was indeed "mad."[1] The problem is that rules, as the philosopher Ludwig Wittgenstein pointed out (and this includes meta-rules), do not contain the rules of their own application.[2] No rule can be applied without doubts about how to apply it in unanticipated circumstances. The rules that gravitational wave scientists impose upon themselves in their determination to avoid accusations of bias will always contain ambiguities. Scientists like to believe there is a set of statistical procedures that, once programmed into their community, will ensure the validity of their procedures and remove the need for human-like decisions. But actually, statistics are just human decisions in mathematical clothing. Every now and again, this causes the kind of trouble that the airplane event exemplifies.[3]

1. Franklin (private communication, October 2009) points out that this situation is not unprecedented. He quotes a high-energy physicist discussing the setting of upper limits as follows: "One ultimately *should* look at events in the signal regions—after all cuts have been fixed—to check whether they are due to some trivial background or instrumental problem such as the high voltage having been tripped off. If such events can be attributed to such sources, then it makes more sense to cut them and set a biased but meaningful limit rather than leave them and set an unbiased but not useful limit. . . ." (The quotation is from a report by A. J. Schwartz, entitled "Why Do a Blind Analysis?" and circulated at Princeton University in 1995.)

2. Wittgenstein 1953.

3. I like to collect examples of moments when even the most long-established rules do not cover new circumstances. One example happened during the World 20/20 cricket tournament held in the UK in June 2009—I saw it on television. In cricket, six "runs" are scored if the batsman hits a ball in the air over or onto the boundary rope or if it is caught by a fielder who touches the rope at the same time (like a home-run in baseball, but the boundary is marked by a rope on the grass and touching not clearing is the criterion). A batsman hit the ball for a potential "6" but a fielder standing just inside the rope leapt up and diverted its flight sharply upward though it was still going to land outside the rope. The fielder then ran outside the rope, leapt off the ground again to push the ball back inside the rope before it fell, and then ran in, picked up the ball, and threw it back, limiting the runs to (I believe) three. At no time was the fielder in contact with the ground or rope and the ball at the same time. I was sure it was a "6" but the umpires counted the ball as not having crossed the rope. If the umpires were right, it seems to establish a new rule allowing fielders to stand outside the field of play—defined by the rope—and, provided they leap off the ground, push potential six-hits, back into the field before they land.

The Equinox Event Breaks the Statistical Rules

The Equinox Event was discovered, as the note that alerted me indicates, by "one of our online searches." An "online search" is, again, something that breaks the rules of the strict division between playtime and data analysis proper, and, again, it is a case where it would be "mad" not to break the rule. For some reason, though, I understand, not without some argument, this institutionalized, and continuing instance, of meta-rule-breaking, which in scientific terms is far more dangerous than the airplane event, has been accepted—that's sociology.

An online search is a real-time study of the data which alerts the collaboration to any coincident event that stands out sufficiently under a rough data analysis to be worth further investigation. As can be seen, it comprises an analysis of main-body data from well before statistical procedures have been frozen. It subverts the effective "blinding," which the split between playtime and serious analysis time is meant to accomplish: once an online search has alerted the collaboration to the possibility of an event, only vigilance can prevent post hoc-ery. But the very idea of blinding, and the very idea of freezing the protocol after playtime, announces the fact that vigilance is not considered strong enough to counter any insidious desire to find a result.

The reason it would be "mad" not to have an online search is that gravitational wave events may be correlated with electromagnetic events such as gamma ray bursts, X-rays, or the bursts of visible light associated with a supernova. When a putative gravitational wave event happens that is "loud" enough to make itself felt even before the immensely complex statistical analysis that follows the "opening of the box," the gravitational-wave scientists want to be in a position to request astronomers to point their instruments in the direction from which it might have come and look for a correlated signal. This has to be done as close to the time of occurrence as possible. In any case, if the event is that outstanding, it might be the first discovery, and analysts would be "mad" if they did not want to start examining it closely as soon as possible.

Many of the dilemmas discussed in this last passage, including that of the online search, are brought out in this interview with a senior member of the collaboration:

> Collins: I hear in this analysis [of the Equinox Event] that you've been looking at this event for a while and trying different cuts on it.

Respondent: Yeah—and that's sort of the impurity of it all. . . . But what we're not doing is changing any of the thresholds, we're not changing any of the operations, any of the lines of code, we're asking questions that we consider "post-processing." However, some people will tell you that your whole procedure from beginning to end must be set up in advance, and the whole concept of post-processing, according to these folks, destroys the statistical purity.

Collins: So you'll get in the neck from some people.

Respondent: Yes—like for example Z. And he's not a fool. And his point is—I think I can represent it properly—is that in the end—of course you have to make all these wise choices about how you analyze the data—but in the end the only thing that distinguishes an event from noise is its degree of statistical improbability of having occurred from noise alone. And . . . that's a huge part of the truth, or it's the whole truth. He thinks it's the whole truth, I think it's a huge part of the truth but not the whole truth. But that's why the thing that everyone wants to show you is: "Here's the histogram of all of random events and see how far to the right this is." . . . So anything you do after you've seen it that either makes it seem more or less probable—that you can't justify as being a mechanical application of what I ought to have done before—at that point you do in fact lose the ability to say how unusual the event is.

Collins: So you said that's a large part of the truth, but what's the other part of the truth?

Respondent: There are huge ideological disputes within the LIGO Scientific Collaboration. . . . The other part is when we say we are doing follow-ups. I think Z is completely allergic to the notion that you should have something that you call a follow-up. You should define your pipeline and then you should open the [box] and write the paper, whereas—and I think this is something where S has been a great partisan—is that you do all this stuff but that's just the beginning—then you've got to see if it makes sense. And you've got to see if there's any signature in the data and look. . . .

There's a classic example in the cosmic background radiation. It's not a tragedy, like the Weber story was a tragedy, but it's sort of a well-known mistake in deciding and announcing to the world that you were . . . Someone wanted to measure the spectrum of the cosmic background radiation at the short wavelength, high frequency end of the spectrum. . . . Paul Richards, the guy who was doing it, was afraid he was going to force it to be a really beautiful black body in his data analysis. And he said, "No I'm not, I'm going to design the instrument, measure it, check everything,

> set up my data analysis pipeline and publish whatever spectrum I get." And he did—he stuck to his guns—but he also realized that the spectrum was wrong, after the fact.—It's like the airplane event!—And people make blunders, and so S's point of view is that we're inventing a new kind of measurement, we're not going to be wise in advance, but we have expertise, and we shouldn't force ourselves to abandon the ability to use our expertise. And there surely is a lot that's right with it, but once you start doing that you lose the ability to quantify how unusual the event was because you can never reproduce the chain of decisions you would have made in other cases. And so we're kind of in the soup that there is something about Z's statistical purity that is correct and yet it's inapplicable most of the time. . . .
>
> These things are delicate, and we know the classic case to either tune to make something go away or tune to make it seem bigger—different people want different things in the secret recesses of their mind.

The whole issue is expressed from a different point of view by an analyst who had been part of a team accused of over-interpreting their data. It seems in the discussion set out below that he was expressing regret at what had happened. At the same time, the conversation brings out both the unconscious desire of the analyst who, as he puts it, "falls in love" with a finding, which implies that nothing should be touched after analysis procedures have been frozen, while at the same time it brings out the need to apply experience retrospectively since not everything can be anticipated in advance. My intervention is that of a "devil's advocate."

> Q: When you have to say if something that has been detected is a real signal, then you must interface with people who remain cold. They must have experience, experimental experience also, because sometimes there are effects that you cannot imagine and only years and years of experience will teach you how many bizarre things may happen to the detector, things which are unpredictable and things which people sitting every day in front of a computer cannot even imagine. A data analyst who has a possible candidate—who studies a possible candidate—can fall in love with this data. Because the data matches some prejudice or something that has been studied already . . . perhaps the first black hole ringdown. Then you feel yourself at the center of history, at the center of the universe at that moment. And in a way there is something—you can feel that it cannot be wrong. It cannot be by chance that at this moment you have this data in front of you. So you must face—you must [try?] your

work, and your suggestions—and make a report of the result of your work to someone who has not been through this process and who hears all the details for the first time.

Collins: But surely if the data analysis group is doing its job properly and doing the statistics properly only one answer can come out?

Q: Data analysis work is not so simple. It is not pushing, first, the red button and then the yellow one then the black and something comes out. There is some role of the human being in the process.

The First Equinox Telecon

My notes from the telecon, when I first heard the Equinox Event (EE) discussed, intimate that I was a little disappointed. The whole thing was very low key. Routine business was dealt with and the first mention of the EE came more than twenty minutes into the discussion. Only after more than an hour did someone say, "[It] looks like a very good candidate."

I suspect that one reason that the telecon was a low-key affair was that the scientists wanted to be professional—they knew that fair data analysis requires vigilance and that too much excitement might compromise it. So things were done in an orderly fashion, allowing the data to "speak for itself," if that was what it was going to do.

A second reason was that everyone knew that even if this did turn out to be a reportable event, there was a good chance that it had been deliberately inserted into the data stream as part of the blind injection challenge; there was no point in getting too excited about something that might turn out to be an injected artifact.

The third reason for my disappointment, as a listener-in on an early discussion of what might just be the first discovery of a gravitational wave, was that the banal nature of the event was being revealed. It simply was not an "event" in the sense that the word is normally used. It was not a sudden explosion, or the appearance of a comet, or a weird freak of the weather, or a Moon-landing, or a new speed record. It was a slightly unusual concatenation of numbers in a torrent of numbers. Without a computer continually monitoring the stream of numbers, there would be nothing there. It was all statistics. My notes taken at the time express it thus:

> I think the weirdness is that there is nothing to "see," just a load of carefully worked out statistical procedures (I guess this is how it is in high energy [physics] too). There is something unsatisfying about the "construction" of a signal from a lot of numbers and procedures. It is something to do with

> being faced right up to the fragility of the inference—it is not just a philosophical nicety but a visceral thing. You feel the fragility—it is just numbers.

Still, they were interesting numbers, and as weekly telecon followed weekly telecon their meaning became steadily more defined.

Two types of reality-transformation were applied to the numbers. First, there was an ongoing statistical examination. How did the numbers look under the different analysis pipelines that the Burst Group operated? And had the Inspiral Group seen anything at the same time? Second, there was the "Detection Checklist" to be worked through.

Sigma Values and the Language of False Alarm Rates

As we have seen, in exploratory physics signals are weak and can only be recognized statistically. As explained, a "discovery" consists of some combination of events, which makes its presence felt as numbers manipulated by a computer program, that is *so unlikely* to have arisen as a result of chance that they have to be counted as a signal—a systematic intervention by nature. The most widely used convention for expressing the unlikelihood of something being due to chance is in terms of a number of "sigmas" or "standard deviations." The theory is that there is smooth distribution of noise in the output of the apparatus so that, if plotted on a graph with size of noise along the bottom axis and number of noises of that size on the vertical axis, the result would be bell-shaped. That is to say, there would be a lot of small noises which would give a high peak around the central zero point of the bottom axis, slightly fewer a little way to the positive and negative sides of the zero point, and fewer and fewer noises as one moves away in a negative and positive direction. At the "tails" of the distribution—the bottom of the bell—there are only a very few big noises well away from the center point. If the noise is "well behaved"—which is to say it follows the right mathematical function—its distance from the center point can be expressed in terms of standard deviations. If the curve has a "Gaussian" profile, then roughly 68 percent of the noises are found within one standard deviation of the zero point, 95 percent within 2 sigma, 99.7 percent within 3 sigma, 99.99 percent within 4 sigma, 99.9999 percent within 5 sigma and so on. One can say, then, that if some signal lies beyond a distance of 2 sigma from the center there are only 5 chances in 100 that it could have been due to noise; if it lies outside 3 sigma, there are only 3 chances in a 1,000 that it could have been due to noise, and so forth; a

5 sigma result would be equivalent to one chance in 1,000,000 that the event would have been caused by smoothly distributed noise. Different sciences have different standards as to what counts as a sufficiently large sigma value to make something count is being so unlikely to be due to noise that it should count as a discovery. Exactly what this standard should be plays a large part in this story.

In gravitational wave detection there is also another way of expressing the unlikelihood that a signal could really be noise. It is also sometimes said that "this effect could arise by chance only once in so many years." Why is this, and how do the two methods of expression relate to each other?

In the case of the LIGO interferometers, the noise plot has too many large noises to fit the smooth mathematical model. These "glitches" (see below) give the distribution a pathologically "long tail"; there are too many big noises out at the extremes of the distribution. Therefore the calculation that gives rise to a statement in terms of sigma cannot be done because the model does not fit.

What is done instead is that the "false alarm rate" is directly measured over the coincidence data produced from the time slides. Because the set of data is limited to what can be generated by using the time-slide technique on the output of the detectors, rather than being the infinite set implied by the existence of a smooth mathematical relationship, there is less certainty about how right it is; it becomes hard to know how much confidence one should have in one's statement of confidence. Still, a likelihood of any one event being due to chance can be induced from whatever data is generated by the time slides. In this case the result was that a coincidence of this amplitude and degree of coherence would turn up by chance roughly once every twenty-six years.

Now, given that the Equinox Event was a three detector coincidence, and the three detectors were all up and running in "science mode" for a total of 0.6 of a year during the second year of the S5 run, which was the basis of the analysis, it means that a single time slide generates 0.6 of a year's worth of spurious coincidences.[4] In fact, one thousand time slides were carried out—each with a different delay. This produced six hundred year's worth of spurious coincidences. Around twenty spurious coincidences that looked as convincing as the Equinox Event were found in the

4. I am not 100 percent certain of the exact choices that were made, but they make no difference to the principle and very little difference to the numbers.

six hundred years, and that means that such a spurious coincidence will show up about once every twenty-six years.

The one chance in twenty-six years can be converted to say that there was 0.6/26 = 0.023, or 2.3 percent likelihood of such an event turning up by chance over the course of the run itself. This probability can then be translated into the sigma level that would be associated with it had it actually emerged from the calculation based on a smooth mathematical model. This language is familiar to physicists and can help them decide whether they "should" take it seriously or not. The 2.3 percent would, under these circumstances, be associated with a roughly 2.5 sigma result. Physicists can, then, refer to the Equinox Event as a 2.5 sigma event (even though the figure may not be accurate) when they need a rough indication of how the result compares with other results in the same field or the confidence typically invoked in other fields. A 5 sigma result, which is the standard for contemporary high-energy physics, would be equivalent in the S5 run of an event that would occur no more than 0.6 times in a million years.

In passing it should be mentioned that the Inspiral Group had seen something in the data that was not incompatible with the Burst Group's Event, but it was not enough to reinforce their claim and would not even have been noticed if the Burst Group had not seen it first.

The Detection Checklist

Of central importance to any potential announcement is the application of the Detection Checklist. This is a long list of hoops that a candidate gravitational wave has to jump through if it is to survive. Each analysis group develops its own checklist. The Burst Group's version as of October 2007, with its seventy-three hoops, is shown in condensed form in Appendix 1. An italicized remark indicates that a task had been completed by the date in question. A more complete version of the list would show who was assigned future tasks, who had completed finished tasks, and what were their sources of evidence, with hyperlinks as appropriate. Though this checklist is six pages long, it is worth reading through quickly to get a sense of what is required to make a credible detection claim. The groups do not allow themselves to make facile claims, and the degree of caution revealed by the checklist process seems appropriate.

Within the Burst Group, interest in the event increased in succeeding telecons as it continued to jump through the checklist hoops without any obvious failures. But, that said, the interest was not very great and not everyone felt it. I think I was more excited than most of the scientists. I decided I would try to test just how much interest there was by looking at the way the news was spreading. First, however, I had to make sure I wasn't about to give away any secrets.

There is a degree of rivalry between the analysis groups; each of the four groups would like to be the first to see a signal. On the other hand, rivalry is muted because, mostly, the groups are looking for different things. In the case of the Burst Group and the Inspiral Group there is an overlap.

The Inspiral Group is looking for the typical "chirp" signature of the final seconds of the inspiral of a binary star system. It does this by matching thousands of templates against the data stream. Matching templates in this way should give a big advantage in detecting a candidate signal, because true signals that match the template will stand up above the noise. The way the process works is that, when there is a correlation between signal and template, the template "lights up," as it were—suggesting that "something that matches this template has been detected."[5] But, as already intimated, in the real world nothing is so straightforward. Sometimes the detector noise will conspire to highlight a template falsely. The Inspiral group applies a second test, the chi-square test, to see if the data really does match the template well. If the chi-square test is applied too loosely, noise fluctuations will still sneak through and highlight templates falsely. This was a worry expressed in an email circulated by one commentator before the techniques were fully refined:

> I was struck by the fact that almost all of the loudest events that the Follow Up team examined were loud events that one could tell BY EYE were not binary coalescence signals. Why is it that the CBC pipeline keeps such events through to the very last step of the pipeline? Isn't it possible for a matched filter search to tell whether a putative signal actually looks like one of its templates? . . .
>
> How is it that these junky . . . signals pass the various signal-based vetoes? Why are the thresholds set so loose that this can happen? . . .
>
> Is it that the group is [so] eager not to lose a single possible detection no matter from what weird corner of parameter space. . . ? Isn't there a

5. My house overlooks Cardiff Bay, and, for amusement, I used to log all the new ships that entered the docks, trying to read the names with the aid of powerful binoculars. The angles or the weather was often such that reading a name was a matter of "extracting signal from noise," just as with gravitational wave detection. It was striking that if the names were written in English—that is, if I had a set of templates to match them against—it was far easier to read them (to initially decipher and then confirm what they were) than if they were written in Russian or Greek script—where I had no template.

> danger that a genuine but weak signal will be lost in the flood of junky loud signals? . . . [Or is it that in some cases] the group has abandoned a matched filter search without admitting it? . . .
>
> We'll want our readers to hear the "ring of truth" in our detection claim. Inability to reject flagrant junk signals will call any detection claim into question.

On the other hand, as the third paragraph of the email implies, if the chi-square test is applied too tightly, and if the templates are not accurate predictions of the true waveforms—and remember there are only a limited number of templates to be applied to an almost infinite number of possibilities—it will also suppress the signals. For some binary star systems, the predictions (templates) are not sufficiently accurate. There are, then, many subtleties in "tuning" the Inspiral search.

The Burst Group, however, is just looking for a coincidence between signals of no particular form, and it could be that they will spot an inspiral which the Inspiral Group will miss for some of the reasons described above. This could be embarrassing, and at least one bit of gossip that later came to me implied that the Inspiral Group felt somewhat shamed by the fact that the Burst Group had seen something that might be an inspiral and they had not. Given this potential rivalry, I had to make sure that the Burst Group did not mind my discussing "their signal" with members of the Inspiral Group. At the same time, from a different source I heard that the Inspiral Group had their own different detection candidate. I made email inquiries of a leading member of the Burst Group about whether I was free to talk to the different groups about the other's work and obtained the following reply, which allowed me to go ahead with my inquiries:

> Although we in the burst group would be pleased to "see it first," we keep no secrets within the LSC-Virgo sphere—at least several members of the inspiral group already know about it. In fact, one of our follow-up checks is to see whether the same event candidate is found by other searches, and in the end whatever statement we can make about the event candidate should reflect having looked at it in as many ways as we can.

By this time I was keeping a running log of everything that was happening—notes on telecons, copies of emails, interview transcripts, and so forth. Here is an extract from my running log made in early October 2007:

I then telephone "A" (burst group) and "B" (inspiral group), to talk about these events. Both telephone calls are recorded.

Neither of them is very excited. Neither of them knows much about the other group's event (B knows nothing at all). A does not much care about either event—he says of the Oct 06 (Inspiral) event that they get something like this every couple of months. (B says it is twice a month.) An SNR [Signal to Noise Ratio] of around 6 [which the Inspiral Group have for their event] is barely different to noise.

A says he is not very excited about the Sep 21 [Equinox] event because it is right at 100 Hz which is where all the glitches are anyway—he obviously think it is a glitch.

B is surprised not to have heard of the event since he had lunch with A yesterday, but I explain that A does not think much of it.

In sum, though I am excited, that's my problem. These guys are too busy with their teaching or whatever to think it worthwhile to investigate even the event pertaining to their own group, never mind that in the other's group. But, of course the 21–09 event will get to the Inspiral Group in due course as a result of normal working down the detection checklist.

In sum, neither of these has caused enough excitement to make anyone change their normal routine—not even to send a group email around.

Note that I have taken an active part in telling at least one member of the inspiral group that the burst group has something. . . .

The point about this episode of fieldwork is that it tells me that my excitement is not shared. It is not shared because in the case of the October event it's just "same old same old." No one believes these things are GWs—though any one of them could be—in fact there is a very good chance that the first real GW will be close to the noise.

In the case of the burst group, at least for A, it is something similar. The thing is in that glitchy area. . . .

So there is this problem, which one can sense in the very lethargy (I should say, lack of much special activity—the item was in the middle of the burst group agenda, even!) that is going to make it hard to actually agree that a GW has been seen. Because hundreds have been nearly seen. How is the epistemological break going to be engendered?

And it is going to be an anti-climax if it is just statistics, or failure to find an alternative explanation. BUT THIS VERY LACK OF EXCITEMENT IS A KIND OF TRIBUTE TO SCIENCE—50 YEARS TO SEE NOTHING EXCITING—MARVELLOUS AS A TYPE OF ACTIVITY. WHAT AN INDICATOR OF THE PURITY OF DEDICATION. [Capitals in original notebook entry.]

Though this extract is from notes taken very early on in the process, it captures the whole "mindset'" dilemma in a nutshell. Few people were particularly excited about the Equinox Event (the Burst Group's potential event) because it looked just like lots of things that have happened before and it like much of the noise that is found in the detector anyway in the region of 100 Hz (cycles per second), which is where it had been found (a point that will turn out to be of great importance in the argument that will come later). The science of gravitational wave detection, as I note in the capitalized passage, has become the science of showing that potential signals are not signals—a monk-like activity which places integrity and dedication to duty far above reward—something admirable in itself. So unexcited were people that they were not even telling others in the wider collaboration about it—not even over lunch! I found myself telling some of the other scientists about it and having to explain that the fact that I am the one who is acting as the conduit does not imply anything sinister.

And yet, there is a good chance that the first gravitational wave to be detected will be a marginal event—something that does look like a noise or a glitch. There is a good chance that the first event will be like this, if people are willing to see such an event at all, simply because the distribution of the sources in the sky means that there are likely to be many more at the outer edge of the range of the detectors than much closer in. "Much closer in" captures a much smaller volume of space than "right out at the edge." The number of sources that can be seen at different distances increases as the cube of the distance, so there are one-thousand times as many sources at a range of 15 Mpc—which, in Initial LIGO, is at the edge of detectability for a standard source—than there are at 1.5 Mpc, where things should be easily detectable. In the case of one of these distant sources, only the long drawn-out application of refined statistics and the checklist will extract it from the background. It is going to be unexciting—just statistical manipulations. As I note sometime later in my running log:

> So that is why it is all so depressing: a real event is just the outcome of messing around with lot of statistics and saying "this is unusual." No atom bombs explode, no power is generated, nothing new at all can be done!!!!

How, then, is the collaboration going to shift people from this instinct of dismissing potential signals, which can itself be felt to be a kind of holy duty, expressing the utmost purity of intention, to seeing a potential event as a real event that deserves all the enhancement it can get outside of post hoc statistical massage? That is the mindset problem.

Of course, the fact that people are going through the checklist, and one of two of them are finding the Equinox Event increasingly interesting, shows that the mindset problem is not necessarily fatal. If data analysis was a purely algorithmical process without any input from human judgment, it *couldn't* be fatal. But there is always judgment involved, as the airplane event and lots more to be discussed illustrate. So the mindset problem could be fatal when it comes to those marginal decisions that do involve judgment.

As it happens, as events unfolded the Equinox Event began to gain salience. There is something you can *do* with such a potential event—discuss it. As sociologists know, discussing things makes them real. The next meeting of the LSC-Virgo collaboration would be in Hannover, and the Equinox Event was going to be discussed. The build-up is indicated by the following email alerting the "Detection Committee," the committee charged with overseeing the announcement of any event and deciding whether to take it to the collaboration as something that should move forward to publication:

> To: the LSC Detection Committee: . . . October 11, 2007:
> A candidate event has been detected by the burst group. None of the search groups are ready to present a case to the Detection Committee but there is knowledge throughout LSC that something is afoot. The burst group is considering giving a brief presentation of their activities regarding this event at the forthcoming Hannover LSC meeting without giving conclusions. It would be a good idea for the Detection Committee to meet to discuss strategy and to formulate the questions that we believe will need to addressed. Suggest that we have a telephone meeting. . . .

When the meeting took place there were immensely long discussions over procedure and exactly how the event was to be discussed. It was concluded that, to keep the excitement damped down, it should be spoken of as just a run-of-the-mill "loudest event" of the sort that is bound to appear in every analysis run. In my notes I express myself astonished that the scientists should be so cautious—the announcement was going only to the collaboration, itself an organization with very restricted membership, my presence in which is an anomaly—though, perhaps, the organization was large enough for leaks to be a concern. As it happens, however, the name "Equinox Event" was soon to become the standard reference, making any attempt to treat it as just an ordinary loud event futile.

Just before the Hannover meeting there was another Burst Group telecon in which the likelihood of the event being due to correlated noise was estimated using three different procedures, producing results of 1 every 100 days, 1 every 2 years, and 1 every 50 years. Extra significance can come from analyzing the data in ways that take more of its properties, such as the energy profile, into account when working out how like each other are the signals in separate detectors. Techniques can also be applied to eliminate glitches with known causes, thus making "the event" stand out higher above the background. The latter method is where the danger of "tuning to the signal" might arise.

The Hannover Meeting

The high point of the Hannover meeting was going to be the discussion of the Equinox Event. But the Inspiral Group seemed to steal a bit of the Burst Group's thunder with an announcement of their own. Apparently, the Inspiral Group, quite independently of anyone else, had decided to run their own blind injections on themselves. They had done this, as they were at pains to stress, without causing any potential trouble to anyone else.

It is the case that the detectors are subject to a continual stream of artificial "signal" injections, known as "hardware injections," which are used to monitor the sensitivity of the devices and the analytic procedures. These false injections cause no trouble because they are not secret—the very point is that the groups know what has been injected when, so that they can check to see that their pipelines were sensitive enough to detect the injections had they been real. The Inspiral Group had simply elected to keep a couple of these regular injections secret from themselves, and they were able to announce at the Hannover meeting that they had found something in the data. I must say I found it puzzling that the group had told the Hannover meeting about this before opening their own private envelope, so that they could also tell the meeting, there and then, whether they had found something real or not. But others said that the Inspiral Group exercise was a good one and the analysts were entitled to their fun. Others said that the exercise was not as useful as the "proper" blind injections because the regular hardware injections were perfectly formed, so it was obvious they were a false not a real signal, which would always be dirty in one way or another. There was no reason to make the calibrations dirty, whereas the true blind injections would be of dirty signals so as to present a more realistic challenge. Anyway, the Inspiral Group's self-imposed blind injections acted as "hors d'oeuvre" to the Burst Group's discussion.

The Burst Group went through the checklist. They explained that though the Equinox Event may not stand out very much in terms of just being a coincidence, when a test for the coherence of the waveform in the detectors was applied, it gained far more significance. A coherence test is always going to be a reasonable thing to do on any coincidence and was planned before the box was opened. The Inspiral Group also reported that they had something of low significance that matched the Burst Group finding but that it did not show up strongly in an inspiral search. The Burst Group spokesperson concluded by explaining that nothing they had seen was "ready to go out the door" and that they were merely reporting work in progress. Nevertheless, they said that, if nothing was found to argue that this was an artifact, they would be expecting to take the result to the Detection Committee.

3 Resistance to Discovery

The driving idea behind the blind injection challenge was to do some social and psychological engineering—to get the community out of its negative mindset. Knowing that there might be blind injections in the data, the scientists had to be ready to see something, not just reject everything. Before going on to explain this in more detail, let me make clear the truth-status of the claim that *I* have just made about the idea behind the challenge. This rationale also covers the status of the claims made in the Introduction, such as: "the strongest pressure for its [eLIGO's] completion came from scientists who believed doubling the sensitivity would be the key to the initial detection of gravitational waves."

The status of these claims is that they are what I heard people saying as I "hung around" the community, acting as a quasi-member of it and acquiring "interactional expertise."[1] Now, a claim of this sort does not prove that these were the motivations that informed everyone. I was hearing snatches of conversation and explanation that fitted into my continually developing understanding of how the community in which I was embedded was working. I made no attempt to do a representative survey to find out how

1. The concept if interactional expertise is developed in Collins and Evans 2007. It is expertise gained through deep immersion in a technical discourse in the absence of practical ability. Experimental tests indicate that it enables sound technical judgments; it is the crucial expertise used by managers when they take on new technical projects (Collins and Sanders, 2007).

many people thought "this" and how many people thought "that." I didn't do this primarily because I do not think surveys are as revealing as hanging around, conversing, and interacting as much as possible, while trying, as far as possible, to become like a member of the group one is studying. I have nothing against surveys as an additional source of information and have tried to use them in this way from time to time, but I do not do it often because they are unreliable and can be counterproductive.[2] They are unreliable because few people answer them, and they can be counterproductive because they distance one from the community. Your friends learn about you from acquaintanceship, not from getting you to fill in questionnaires about your eating, drinking, and reading habits—at least not beyond the introductory round of Internet-dating. Finally, I am after something more than the aggregate of individuals' motivations. I am looking for the "spirit" of the discussions; this is expressed by what can and cannot be legitimately stated in public. I am looking for what has been elsewhere called "the development of formative intentions" or legitimate "vocabularies of motive" rather than trying to work out what is in a set of different individuals' heads. Working out what is in individuals' heads is the extremely difficult and demonstrably fallible job of law courts and the like. I am not equipped to do it. My approach, of treating the kind of things people say to me as indicating what is say-able at that time, rather than looking for their "true" motives (as though we even know our own motives with certainty), I call the Anti-Forensic Principle. The Anti-Forensic Principle indicates that sociology is concerned with the nature and "logic" of cultures—in this case developing local cultures—not the guilt, innocence, or motive of any specific person. One acquires an understanding of the nature and logic of a culture by chatting in coffee rooms and corridors, not by doing surveys.[3]

It is the case, however, that not long after the blind injection challenge was put in place, various scientists offered explanations of why it was a good thing that were different or additional to the "changing the mindset" explanation that I have just put forward. Some scientists indicated that these other reasons were what motivated the blind injection in the first place. I can't be sure that these reasons were not the initial motivating factor for at least some these people. It does not matter: a good proportion

2. An example is presented below.

3. The investigator's interactional expertise (Collins and Evans 2007) provides the warrant for claims and questions about the science expressed here in the first person or simply stated without further justification. For "formative intentions" see Collins and Kusch 1998. For "vocabularies of motive" see C. Wright Mills 1940. For the first invocation of the Anti-Forensic Principle, see *Gravity's Shadow* (412).

of the community thought it was being done to change the mindset of the community, and this made a lot of sense because it was felt that the mindset was in need of changing. That the challenge served other purposes too, even if these were not the driving force, is also beyond doubt; these other purposes will be explained in due course, but first the "mindset problem" will be examined.

How does a mindset become established and get transmitted to new generations of a community? In large part it is through relating "myths" or telling "war stories." The term "myth," used in this context, does not mean a false account. Rather it refers to the first part of the definition as found in my *Chamber's Dictionary:* "an ancient traditional story of gods or heroes, especially one offering an explanation of some fact or phenomenon; a story with a veiled meaning." Here there are no gods, nor even heroes, but only antiheroes, and the meaning is not veiled. The subjects are Joe Weber and/or "the Italians," and the stories of their wrongdoing are circulated and recalled with words such as "to analyze data like that would be to risk doing what Joe Weber did," or "to claim that would be to make statements as irresponsible as those of "the Italians." You can hear those phrases repeated in the corridors of interferometer meetings whenever some judgment is being made about data analysis. Recalling those stories provides a guideline for new scientists on how they are to act and reinforces the views of the scientists who already know how to act. To act in any other way would be to violate a taboo and act like a scientific degenerate.

Joe Weber and Statistics

It is now widely believed, though never decisively proved, that Weber managed to get the results he got by manipulating his statistics retrospectively.[4] Weber had a tendency to "tune to a signal." As I argue in *Gravity's Shadow*, if he did do this, Weber would have been doing the thing that served him well when he commanded a submarine chaser in the Second World War. When you are trying to find a submarine, you turn the dials of the detection apparatus backward and forward as you tune in to the telltale echo. And it is vital to find the submarine if it is there. If it is there and you fail to find it, some of your compatriots will likely die. If you "find it" and it is not there, nothing worse will happen than some waste of time and possibly ammunition. In submarine chasing it is much better to choose a strategy

4. There were many other ways in which Weber could have gone wrong outside of poor statistics; it is just that the statistical explanation has become standard.

that will result in false positives than false negatives, and the skillful commander will find the submarine even if it is at the cost of quite a few false alarms.

But tuning to a signal is dangerous in physics, or, at least, it is dangerous in exploratory physics. Suppose a standard has been set such that a set a numbers reflecting some kind of potential discovery will only be counted as a discovery if it would arise, as a result of chance, only one time in one thousand searches. The trouble with tuning to the signal is that each adjustment of the tuning dial represents a new search. Thus, if you twiddle the dial 100 times before you find the 1 in 1,000 you have done a hundred searches. An event that is only likely to turn up one time in one thousand in one search is likely to turn up one time in ten in a collection of a hundred searches. Thus, if you report only one of those searches, the event will look like a discovery, but it does not meet the standard of a discovery at all.

If the meaning of the last paragraph is not immediately obvious, it is worth going back to it until it is because a great deal of what follows turns on its logic. A major claim of this book, remember, is that, however much a discovery is presented as a unitary reading, it is really just an eddy—a small area of seemingly still water—in the onrushing stream of history. One can see, if one follows the logic of the last paragraph, that a paper that claims a 1 in 1,000 value discovery—something which to follow the metaphor, is an undisturbed patch of water in the stream—might well be swirling around at a speed appropriate for a one in ten event. It all depends on what happened immediately upstream of the eddy, such as whether the water was disturbed by lots of twiddling of dials. Notice that how much dial twiddling took place upstream is not generally mentioned in the scientific paper which reports the result—the paper simply reports the result of the final calculation of unlikelihood, and that is why it can be so misleading. To understand the true meaning of a published statistical claim you have to be a cross between a historian and a detective. And, as we will see later, you have to be able to read minds too.

The fact that statistical reports, though presented as timeless readings, should actually be read as historical accounts is one of the reasons it is impossible to know for sure if Joe Weber's "results" really were the outcome of what is known as "statistical massage." To know that would be to know exactly what Joe Weber did with his dials, and we don't. The fact that it is widely believed, to the point where it has become a myth, or regularly told war story, belonging to the field of gravitational wave detection, is, however, a fact with real consequences. It makes the whole community extremely sensitive to what can go wrong. It is certainly a powerful contributor to the mindset and provides one of the resources that scientists can turn to in their debates should they want to count an apparent signal as insufficiently convincing to count as a discovery.

"The Italians"

Not all Italians are "Italians." When someone in the LIGO community refers to "the Italians," he or she has in mind a specific group who ran cryogenic bars and were based in the Italian High Energy physics laboratory widely known by the name of the small town, just outside Rome, in which it is situated: Frascati. As some of the Frascati physicists also held jobs in Rome's universities, they are also known as "The Rome Group." There are about half-a-dozen "Italians." Nowadays, as with Joe Weber, "the Italians" are used for myth-relating and mindset-setting.

Over the last half-decade all the detector groups worldwide have gathered together to share their data with one another. In the early days, each new detector group believed it had achieved the technological breakthrough that would finally achieve the elusive detection, but in each case years of frustrating work proved that the promised levels of sensitivity were not going to be reached any time soon. LIGO was the only group to reach, or nearly reach, their design sensitivity in some approximation to their projected timetable, and even it, in spite of having by far the biggest team, reached the promised landmark about three years late. No other interferometer team has yet come as close to its design sensitivity as LIGO.

LIGO's policy, under the leadership of one-time, and now once more, high-energy physicist, Barry Barish, was to bring the international groups together. The logic of gravitational wave detection means that in the long term data-sharing is inevitable because only by "triangulating" the signals seen on detectors spread across the planet can the direction of a source be located. Each individual detector is almost completely insensitive to the direction from which a wave comes, and it takes four widely separated detectors to pinpoint the source, working from the different times that the signal—assumed to be traveling at the speed of light—impacts on each one. Barish was used to working in international collaborations such as CERN, so, bringing the groups together was a natural thing for him to do.

The Virgo detector, located near Pisa, is the only one which can begin to compete with LIGO, and the initial relationship between the groups was one of rivalry. Virgo has a better suspension which, in principle, should allow it to detect signals of a lower frequency. The spiraling together of a binary system composed of large black holes should emit waves at low frequency for some considerable time; such waves would be invisible to a higher frequency detector, but would make a distinctive waveform easier to spot.[5] There are also many low-frequency pulsars that could not be seen by LIGO and for which the signal would build its signature over time. This gave the Italian-French team some chance of making an independent detection even though they had only one detector of 3 km arm length as opposed to two at 4 km, so that they could not back up a claim by seeing coincident signals between two machines.

Too much rivalry could be fatal for all the groups: suppose LIGO saw an event but Virgo did not? Virgo might insist that this implied that there was something wrong with the LIGO claim. I do not know whether this was part of Barish's, or any one else's, motivation, but in a case like this, to borrow the words of Lyndon B. Johnson, it is better to have everyone inside the tent pissing out than someone outside the tent pissing in.

An agreement between the main groups became easier when it turned out that the specially good low frequency performance of Virgo was not going to be realized until some time after the intended higher frequency performance had been achieved. Indeed, for a long time LIGO had the better low frequency performance. This, with the dawning of the realization that no-one was going to make the first discovery very soon, took the heat out of the rivalry, and by around the end of 2006, Virgo and the LIGO Scientific Collaboration (LSC), which already included GEO 600, agreed to collaborate and hold joint meetings. The collaborating group was named LSC-Virgo. The clumsy name indicates that an institutional merger may not be a complete merger in every sense. As we will see, residual suspicions play a part in the story.[6]

The point is that there are many Italians in LSC-Virgo who have the same attitudes toward the "Italians" as do the members of the LIGO group. Indeed, as compatriots, they may be more embarrassed by what "the Italians" did. The problems of "structural balance" in these relationships are complicated by the facts that one of "the Italians," Eugenio Coccia, has become head of the Gran Sasso Laboratory, one of Italy's leading physics institutions, and that a good number of "the Italians" have now joined the

5. It has become less clear that massive black-hole binary systems near to the end of their lives will have had time to form during the lifetime of the universe, and, to date, no such binary systems have been detected by any means.

6. To use the language of *Gravity's Shadow* (665–67), while the two groups exhibit "system integration" sufficient to give rise to "technical integration," "moral integration" is incomplete.

LSC-Virgo collaboration. The tensions are illustrated by a rather good joke made by one of my friends in LIGO who, echoing a line from *Dr. Strangelove*, remarked to me when he heard that several of "the Italians" were to be present at future meetings of the closely guarded LSC, "but they'll see the Big Board!"[7]

So what did "the Italians" do to become quasi-mythical exemplars of bad behavior? They continued to claim to have seen gravitational waves of a similar energy to those Joe Weber had claimed to see long after Joe Weber was discredited. Worse, in two cases they announced results where the signal was in coincidence with Joe Weber's signals. There were four sets of announcements altogether involving the Rome Group of which two were particularly notorious, with the last being the one most often mentioned today as "a warning to children."

The first occurrence, which attracted almost no attention, was a paper published in 1982 showing correlations between Joe Weber's room-temperature bar and a cryogenic bar run by the Rome Group but located in Geneva. This paper claimed to find a "zero delay excess" of events associated with a 3.6 standard deviation (or 3.6 sigma) level of significance. Remember, a "zero delay excess" means that the two widely separated bars vibrate strongly at the same time more often than they spuriously appear to vibrate at the same time when the two data streams are offset in time. The offset comparisons, or comparisons with delays, or comparisons with time slides—which are three ways of saying the same thing—show what should be expected as a result of coincidences between noise alone—the "background." A finding that there are significantly more genuinely simultaneous coincidences suggests that something external caused that zero-delay excess.

Each time slide will produce a slightly different result for the number of chance coincidences. From these results one can create a histogram that shows how many times different numbers of spurious coincidences show up in the complete set of time slides. One will find that the histogram is peaked at some average level of spurious coincidences, while the further one departs from that average number the fewer slides will there be that exhibit that number of coincidences. As you go right out to the tails of the histogram, which are a long way from the average, there is very little

7. In *Dr. Strangelove*, the "Big Board" is the graphic representation of the location of American warplanes and their attack plans found in the top-secret "War Room." An American general makes the remark when he learns that a Russian is to be invited in to learn the plans as a last chance of averting an accidental nuclear war.

chance of finding slides with those very many or very few chance coincidences. A 3.6 sigma result means that one would expect such a result to be generated by chance coincidences caused by noise alone only about one time in ten thousand—some way toward the tail of the histogram. Attaining a 3.6 sigma result would provide a fantastic level of confidence in psychology or sociology, where 2 sigma, or five chances in one hundred, is regularly accepted, but it would not count as much of a result in contemporary high energy physics, which prefers 5 sigma, which is about one chance in a million. We will return to the choice of confidence level; for the moment it is enough to note that in 1982 "the Italians," along with Joe Weber, published a result that was strong enough to be irritating if anyone had taken any notice of it, but they didn't.

People took much more notice of a 1989 paper, which was again a collaborative effort between the Rome Group and Joe Weber. In this paper they claimed that their respective room-temperature bars, which were the only detectors "on air" at the time, had detected gravitational waves emitted by a supernova—the famous Supernova 1987A—which was visible from Earth and which emitted a flux of neutrinos that was also detected. In that paper they calculated that the energy needed to enable their detectors to see the waves would have needed the total consumption of two thousand solar masses. Although the consumption of this much energy in a supernova was incompatible with either known physics or known astrophysics, they published anyway, exerting the right of experimenters to see theoretically impossible things. Their announced result also caused a fuss because at this time there was an active campaign to fund LIGO; if the result was valid, then the assumptions on which the huge and expensive LIGO were based were wrong. The result reached the science news media but was soon ruled out of court by the majority of the physics community.

The next incident of "bad behavior" was the 1996 announcement, in the form of conference presentations, that the Rome Group and the group in Perth, Australia, led by David Blair, had found suggestive coincidences between their bars. This result never got as far as the journals, but it was discussed at several conferences—because of the determination of Blair, and to the extreme annoyance of the nascent interferometer community. Guido Pizzella, the then leader of the Rome Group, told me that he felt too bruised by the reaction to the Supernova 1987A claims to expose himself to further scorn. In 1996 Blair told me, however:

> I think it's not right . . . because of the fuss that there's been in the past, to deny the data. . . . [W]e were not going to be bullied by people who have

their own agenda. We believed that what we had seen was reasonable and interesting, and that you should tell the story as the story goes—as it unfolds.

Blair made the conference presentations.

Once more the energy calculation showed that, if Perth and Rome were seeing something, the assumptions upon which hundreds of millions of dollars were being spent were flawed. The members of the interferometer community were furious; they felt that every "serious physicist" should accept that to detect gravitational waves, especially in the large numbers required to have a significant "zero-delay excess" in the bars, one had to wait for the interferometers, which were going to be many orders of magnitude more sensitive. From the point of view of the interferometer community, the only purpose that could be served by these announcements was to cause trouble, to remind people of the days of Joe Weber's shameful claims, and to project an even more flaky image of gravitational wave physics to the already skeptical scientific community.

The final incident was when "the Italians" published a paper in November 2002 claiming that, in 2001, they had seen coincidences between two of their own cryogenic bars, one located in Frascati and one located in Geneva. At this time of growing integration in the international community of gravitational wave physics, the publication came as a shock because the paper had been submitted before being seen by the community, which would likely have tried to suppress it.[8] "The Italians" knew they would find no friendly reviewers among the LIGO-centered groups and so went straight to the peer-review process of the journal *Classical and Quantum Gravity*, which published what, on the face of it, looked like an interesting result. Its special interest lay in the fact that the zero-delay excess coincidences were clustered within certain hours during the day and that the particular hours moved around the clock during the course of the year in such a way as to suggest that the source of the energy was related to the Galaxy and not the solar system.

This claim echoed one made by Weber for his own results in the early years. Unfortunately, in the case of Weber the effect disappeared after a while and it dropped from view. The power of this kind of claim lies in the fact that, while it is easy to imagine some spurious source of coincident

8. The flames were fanned when *New Scientist* (9 November 2002) picked up the controversy and, finding that some physicists were unwilling to offer comments to their reporters, wrote an editorial with the tag line "Hushing up scientific controversies is never a smart move."

disturbance correlated with the changing solar day—much creaks and groans as the Earth turns its faces to and away from the Sun, with different things being heated and cooled, tidal forces exerting their effects, and so on—it is very hard to think of anything that is correlated with the sidereal clock. The sidereal clock shifts around in respect to the solar clock as the year passes. As the Earth goes round the Sun in the course of a year, the way it is orientated in respect of the center of the Galaxy at any one time of day goes through a complete cycle. For example, if London faces the center of the Galaxy at noon on a certain date, it will be sideways on to the Galaxy three months later, will face away from the Galaxy three months after that, be sideways on again but the other way round three months after that, and be back to full face again three months after that. If the clustering of signals seen by the pair of bars moves round the clock in the same way, it suggests that something to do with the Galaxy, not the Sun, is causing the clustering, and it is hard to think of what it might be if it is not gravitational waves.

The counterargument, presented as a verbal assault, was that the statistics of this clustering did not stand up to examination. Once more, statistical analysis, it was said, had been converted into wishful thinking. The true statistical significance of the clustering, when the calculations were done properly, was only one standard deviation, and this amounted to nothing. Whether this was exactly true or not depended on a fine subtlety of statistical thinking which will be discussed in chapter 5. Here, however, it is the reaction to the claims that is being explored.

In December 2002 there was a meeting in Kyoto of the Gravitational Wave International Committee (GWIC), which coordinates the various worldwide gravitational wave detection projects, including cryogenic bar detector and interferometer projects. The Rome Group publication was discussed at the Kyoto meeting. The following excerpts from the discussion give the flavor of the debate. The chair, Barry Barish, who was also the director of the LIGO project, introduced the item as follows:

> As we start having results in this community we need to figure out our own guidelines and standards. For those who come from outside of this community, it's fair to say that this community doesn't have a very good reputation—that the past history of having results that haven't stood up hasn't served the community very well, and I think it's important to have credibility and to do things right.
>
> I come from a community in particle physics where there's been a lot of good results, and exciting results, but also wrong results, and that commu-

> nity had developed throughout the years certain ways of presenting results, and vetting results, that may or may not be the model for here but is at least one we should discuss.
>
> There's a couple of things that stimulate this discussion for me. One is we're just getting to the point in LIGO where we're going to have our own results . . . and I'll show you a picture of our present plan but it's kind of developing—of how we go through keeping the results private, how we present them, and in what ways we bring them to publication. And there's a recent publication by, er, er, your group [indicating one of "the Italians" who was present], that raises certain questions, I think, and I want to raise those so we bring them for discussion.
>
> First let me say some ideas and show you one picture from us, which may not be the right ideas but at least I can tell you what the reasons are. They come from some experience in particle physics. In particle physics there's a lot of ideas that when they first come out people object to, and some history that's made when things finally come to publication. There's a certain vetting process that's been healthy. And I propose we need some of that in our own community before we present things. . . .
>
> And in a community that hasn't been fully credible in its past history, it may be very important—you have to build a credibility if people are going to believe results that come out of this field. . . .
>
> As we have results, we really need a way that we all try to follow—not exactly—the same rule, that we try to follow as we go forward to present these results so that we present the best kinds of paper which contain the best information and not debate them after they're in press—I think that's the worst problem is if you debate them after they're in press. Inside our community it makes even things that can be very right—they can be right or wrong—but it makes the whole thing controversial—unnecessarily.

The project director of LIGO, Gary Sanders, remarked:

> Let's say there's a paper comes out and the whole community comes back and says we don't believe it. Let's say that happens. Where would you rather have that in? After the preprint comes out before publication, where the author of the publication has a chance to hear this and think about it, or you submit the publication and hear about it later. Now it's on the record and now it is, as [. . .] described, somewhat divisive in the field. It divides the field into those who are willing perhaps [to] take more risks, to put it in the terms, and those who would like this field to develop as a conversation.

To this the following response [lightly edited for English] was made by a spokesperson for the Rome Group:

> So, first of all let me say something. The publicity—I really did not want the publicity. The journalist at *New Scientist* called me because, she said, of the paper in *CQG*. The paper was written in a way in which we were very careful in the title and the abstract and the final conclusion. It's not a claim—it's not a claim—it's just results. Even the analysis is very simple—I mean just reporting coincidences versus the time. And it describes the procedure of data selection. It is very simple and, let me say, it is very poor. The paper on the statistical analysis will be presented here by Pia Astone at this conference. And now that GWIC has proposed some guidelines we can follow these steps in future.
>
> May I say that in the absence of guidelines from GWIC we followed the guidelines of our group. And the guidelines of our group have always been, like in many other communities, to send the paper to a journal, without any restrictions on referees, expect the referees reports, and then modify the paper or not—depending on the outcome. And then, once the paper is accepted, circulate the paper in the community.
>
> I think this is a procedure of many people. People put a paper on the web only when it is accepted by journals. Not all the people, but many people in the experimental field do that. So this is just to follow the story.

Barish responded:

> But not so much in fields that get picked up by *New Scientist*. That's why particle physics doesn't do it that way. Because the fields that tend to get picked up and made very public have a different—you live in a fishbowl—it's a different kind of environment. So it's true, but I don't think you can identify [those who do what you said as] a high profile community. We're a high profile community.

The Rome Group spokesperson then added:

> Let me say something else. Of course the feeling was that this was not a normal paper because we had the feeling that we had hit something that may be important. So we discussed it in the group—how to proceed. First of all we made a lot of controls . . . and then if we should first send preprints or send the paper directly to the journal. The majority of the people were of the opinion that we should send direct to the journal and receive the

> reports of the referees as a part of the procedure in which the referee plays the role of a sample of the community, let us say. . . .
>
> I have to say that the referee reports were very good, so, if the paper is correct, why not diffuse this information? We make it known: other groups now know that they should look at a time when they have a good orientation to the galactic disk, and we show exactly the time and the date of the events to give the community information. If there are flares, if there are X-rays, if there are gammas at those times, how else can you reach everybody? Of course you can put something of the web but you can also publish. . . .

He remarked later:

> The reason why I personally as a group leader accepted that we should publish the paper was because it was published not as a claim or as a evidence, but it was reported as a study of a coincidence actually taken by the detectors at their best performances and only with this in mind—that it was not a claim, just an *indication* [my emphasis]. Also to declare how we will do the next analysis. So we told before what were the parameters, we made the choice, the selection, the procedure, and then using that procedure we found something that is not a claim but which gives some—let's say, can fix the procedure for looking to the next one and for contributing to the discussion in the community giving the feeling that doing [things] in a certain way one sees something that is unexpected but which can be very important. . . .
>
> I understand perfectly the sensitivity of the field, and for that reason we published the paper as a study of coincidence not as a claim and we didn't look for journalists at all. This was something that happened in spite of our attitude not because of it.

A representative of the non-"Italian," Italian groups summed up some of their worries as follows:

> I think we are discussing something different. This is that this is a high profile field because there is a lot of public expenditure in it. The reason why there is a lot of public expenditure is that this is a forefront field of physics and we all agree that the reputation of the field is very important. . . .

The other Italians, not in the "Italian" group, thus shared the concerns of the LIGO group.

In 2003 a statistical analysis carried out by Sam Finn, which argued that the statistical significance of the Galaxy-related clustering of signals was really only one sigma and therefore counted for nothing, was published by *Classical and Quantum Gravity* in the form of a response to the Rome Group paper. "The Italians" replied to this charge. Their response, however, was not published in the journal, and it found its final resting place in the electronic preprint archive. That, as far as the bulk of the gravitational wave detection community was concerned, was the end of the 2001 coincidences and the 2002 publication. I believe that whether "the Italians" were right or wrong in respect of their findings, the statistical argument was too revealing to be forgotten. It will be looked at in detail in chapter 5.

Discoveries as Binary Events

The charge of weak statistics was not the only one thrown at "the Italians." Another accusation could be heard, but only in the corridors rather than the public sessions. Nevertheless, corridor discussions are equally "formative" of the ethos of the field. What was said was that "the Italians" were acting sneakily in the way they defended themselves. The Rome Group spokesperson insisted that the paper "was published not as a claim or as a evidence but was reported as a study of the coincidence actually taken by the detectors at their best performances and only with this in mind—that it was not a claim, just an *indication*." To the Rome Group this seemed a perfectly reasonable way to defend their actions, but the large proportion of the gravitational wave community thought it dishonorable. In 1995 a senior American scientist had explained this to me in reference to some of the earlier "Italian" results:

> It was mostly at these general relativity meetings. . . . They would give their presentations in such a way that they would lead you, they would show you this data, and they would show you the events, and they would show you some statistics they'd done, but never enough of it so that you could really get your arms around it. And they left you with this tantalizing notion that they could go either way. They either could make a claim that they had detected something if they had wanted to, if they had gone the next step in their presentation, or they could back off and say, "Well, yeah, maybe the statistics isn't good enough." And they left you at that critical juncture. . . . What would happen is that you had to draw the inference. Now that gave

> them the freedom at any point later on, or maybe even . . . to say, "Well, if we choose to say this, we have detected it, or if we choose to interpret that way, we haven't detected it, because the statistics isn't good enough." It was this . . . ambiguity—OK? That got to me—OK?

Scientists worry that a research claim that is open to different interpretations could too easily be used to the researcher's advantage. A publication of this kind that turns out to anticipate a more decisive claim confirming the positive interpretation could grab the Nobel Prize even though the work had not been finished; if the positive interpretation is not confirmed, however, there is no penalty because no mistake has been made—the initial publication was too equivocal.

In the Introduction I argued that discoveries are eddies in the onrushing stream of history. But most physicists do not see it that way. Physicists see a discovery as something that either did or did not happen: the discovery switch can be either "on" of "off" but not in between; the symbol is either a "1" or a "0." And the Nobel Prize is the icon of the digital nature of the discoverer's world. You can win or share a Nobel Prize or not win or share a Nobel Prize, but you cannot sort-of-win or sort-of-share a Nobel Prize: to act as though you *might* have done something that deserves a Nobel is to threaten to wreck the whole "discovery" edifice which has so long supported physical science.

And that is why in 2009, in a hotel restaurant in Arcadia, I could listen to a senior gravitational wave scientist telling me that he did not want anything he published to

> read like that Rome paper from whatever year it was—2001 or whatever—where what was infuriating about that was its coy playing with "we don't really have good evidence but we want you to think maybe." And that paper haunts us. And people draw different lessons from it, but what haunts me, and some other people, is not wanting to be accused of trying to straddle the line of winking at people when we aren't actually able to stake our honor on it. . . . trying to stay on the sober side of coyness and winking.

That is how the "myth" of "the Italians" continues to play its role and construct the ethos of what it is to find something new about the natural world. There is to be no winking. The sentiment will return when we approach the end of the story.

Blind Injections and Their Problems

It was at the Hannover meeting that I first learned, to my surprise—and I discovered that I was not the only one who was taken by surprise—that one of the purposes of blind injections has been retrospectively discovered to be fending off press speculation about unusual activity/excitement in the collaboration and to discount leaks. Even if the collaboration becomes noticeably heated in their analysis of an event, no-one can know if it is a real event or a blind injection until "the envelope" is opened. The "no-one" includes "deep throats," unless they go in for the most mundane kind of cheating by looking at the channels which record the injections—which everyone has vowed not to look at—and journalists (and me) who wouldn't know where to look. Only two people in the collaboration are responsible for the injections, and, crude cheating aside, only they know what their random numbers procedure prompted them to insert and record in the envelope. Even cheating could be avoided by enciphering the channel into which the fake inputs are inserted, but the scientists are worried about going in for this degree of complication in case something stops them from decoding the channel again when the time comes. In any case, even if someone did cheat, it would not be that some incorrect physical fact would be announced to the world; it would just be that the intended social engineering did not work properly.

The social engineering—and this has to be assumed to be the main purpose of the blind injection challenge—was intended to change the mindset of the scientists from non-detection to detection. They had to find the blind injections to demonstrate their ability to find a gravitational wave. It sounds like a great idea, but it turned out to be a lot more complicated than anyone, and that includes me, who can legitimately claim to have some expertise in understanding "social things," had realized it was going to be.

The biggest danger was that the blind injections would have exactly the opposite effect to what was intended. Extracting a signal from noise is a lot of work. If scientists thought that all this work was being wasted on an artifact, they might be still less willing to do it! One scientist put it this way: "All your enthusiasm gets sucked away. . . . It's messing with your head . . . in a complicated way." This was quite a widespread opinion, and another scientist said it had a very real consequence for him:

> I concluded early on that the [Equinox] event was an injection and have not been willing to devote as much urgency in investigating it, as I would

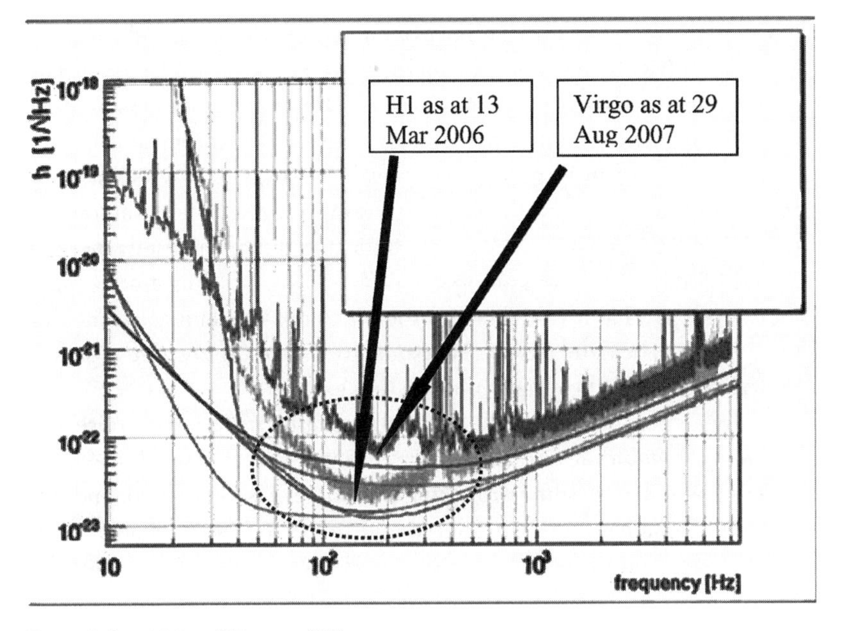

Figure 3. Sensitivity of Virgo and H1

have otherwise. I'm just not willing to drop everything else to hunt down a deliberate false alarm (although I'm glad others are).

The other big problem was that quite a bit of energy went into trying to work out, not whether the event could represent a real signal, but whether it was in fact a blind injection. This was well outside the spirit of the exercise, but if one could work out that it was in fact a blind injection, then it might be more tempting to adopt the approach of the scientist quoted above.

As it happens there were clues that could be used to indicate that the event was more rather than less likely to be a blind injection. First, the whole exercise had not been put into place until quite late in the final two-year/one-year run known as S5, and the event was in the right time period; this made the odds about 3:1 for an injection. Secondly, the collaboration between LIGO and Virgo was relatively new, and it was not possible to inject a corresponding signal into Virgo. Therefore the event had to be injected in the frequency waveband where Virgo was relatively insensitive compared to LIGO, so that it could be seen by LIGO but not by Virgo. Figure 3 shows the sensitivities of Virgo and H1. Sensitivity is indicated (inversely) on the vertical axis (the smaller the strain that can be seen the better) with frequency on the horizontal. The dotted oval indicates the area where LIGO is sensitive enough to see a small event and Virgo is not. As can be seen, this waveband is around 100 Hz or 100 cycles per second.

Restricting the blind injection to within this waveband meant that the absence of a signal in Virgo was not a dead give-away. As it was, for those inclined to think that way, the fact that the event occurred in that low frequency band, though it was not a clear indicator, did substantially increase the odds that the event was an injection. There was a feeling among many of the scientists that it would be just too much of a coincidence if the first gravity wave to be detected just happened to fit into the restricted parameter space demanded by the blind injection exercise.

Another interpretation of the utility of the blind injection exercise, and it is one that would never have occurred to me, is that it would calm the analysts down in case some promising candidate was discovered, and that this would produce more considered and therefore more reliable analyses. That viewpoint is reflected in the conversation which concludes this chapter.

Another problem, also reflected in that conversation, is whether the sequence of events that follow a blind injection can really be the same as that which would be followed in the case of a real event. In other words, where does the process end and at what point is the envelope opened? The longer the envelope stays closed, the more time is wasted going through spurious exercises; the sooner it is opened, the less like a real event analysis is the fake analysis. As it was, the idea was that the envelope would stay closed long enough for the groups to write draft papers as though for publication. It turned out, however, that they wrote only the abstract because working only that out was, as we will see, uncomfortably time-consuming and labor-intensive. More worrying is the fact that collaborations were set up with groups of astronomers looking at the electromagnetic spectrum—X-rays, radio waves, gamma-ray bursts, and so forth. The idea, as mentioned earlier, is that these groups should be alerted so that they can look for bursts in their own wavebands that correlate with gravity waves and that appear to come from the right general direction. But can one waste these other groups' time on what might be an injection? This issue does not seem to have been settled.

The advantages and disadvantages of the blind injection idea are summarized in table 1.

In October 2007 I carried out a small opinion poll to find out what scientists in the Burst Group thought about the blind injection. Not untypically, only five replies came back, but the variation is interestingly large.

Table 1. Advantages and disadvantages of blind injections

UPSIDE	DOWNSIDE
Encourage analysts to work hard because there should be something to find	Discourage analysts from working hard because could be wasting time on an injection rather than an event
The doubters might get egg on their faces if something they say is not worth pursuing turns out to be an injection	It takes the excitement out of the analysis—people just shrug instead of being animated
Discourage analysts from working too hard or too obsessively so work is not hurried but careful and deliberate	Waste valuable analysis and thinking time on artifacts; cause spurious and unnecessary excitement
Help keep a lid on potential findings because they might always be an injection – the outside world can be told this	You can't really go all the way because it wastes more and more time as you go further (eg to the electromagnetic groups)
Enables detection procedures to be rehearsed before the real thing	*Don't look now but a corresponding downside will be discussed at p 135*

Table 2. A little survey—October 2007—on what the Equinox Event might be

It is an injection	It is a signal though it may not justify a discovery claim	It is correlated noise
45	45	10
85	5	10
40	30	30
40	20	40
75	10	15

The replies are shown in table 2. Scientists were asked to report their feelings about the nature of the Equinox Event by assigning probabilities, summing to 100, for each of three possibilities.

Hannover Coffee Discussion

At the Hannover meeting I found myself discussing the event with four of the scientists over coffee. Here is how the conversation went:

> Collins: I'm interested in your best guesses about what's going to happen to this burst event.
>
> A: It will be declared as an injection.
>
> Collins: Yeah, but if it's not declared as an injection?
>
> B: The trouble is it looks like a lot of other things so that's going to make it difficult. But if it's highly statistically significant, we can't throw it out so

we'll have to write a paper. . . . We might not want a detection paper, so we'll have to write a paper that says "we saw something, but we can't say for sure what it is."

A: My feeling is that it'll be an extremely interesting candidate, but I think we will not be publishing results based on few gravitational wave events. If it is only one event, we have to see more of those.

B: We have to publish something, otherwise we would have no burst publication from S5 and that would be the only other option and that would be also . . .

C: Otherwise you're doing an airplane to it and you're throwing it out without any reason to throw it out.

B: Exactly.

D: So I think what will happen is that this decision will be made before we decide whether it's a blind injection or not. If I was a gambling man I would also wind up saying it will probably wind up being a blind injection, but it will be interesting to have to make that decision before we know.

C: I don't think we can know what bias knowing the possibility of a blind injection could be in there has on that decision.

D: Oh it's a huge bias.

B: But the other thing we can't do is go ask the rest of the astrophysical community if they saw something around this time, before we open—well I guess we could do it but it seems crazy to do that—before we open the envelope.

A: No. I think we could do the following thing—"Look, we have got here a program which says that we will do blind injections, and therefore this candidate might be a real event or it might be a blind injection—we don't know that but we would like to find out whether you had any events." . . . That will make them not so excited, but if you do this only once a year. . . .

B: I think it's coloring peoples' actions to have, in fact, blind injections. If there were no blind injections, we would be behaving differently, and maybe the blind injections helps us to behave a little more honestly in some ways.

C: And we're calmer because there are blind injections.

B: But maybe we're not taking it seriously enough—because there are blind injections.

. . .

C: But don't forget, the blind injections do another thing, which is—a big worry—let's say we can't find an explanation—let's say we didn't have the blind injections—we can't find an explanation for this—we don't want to stand behind it as a publication of a gravitational wave—you know, what do we do? Do we throw it out? Do we not publish on S5? The blind injections are going to keep us thinking it's OK to come to a conclusion that we think there's something there in the data because we think there could be something there in the data. You know, if the blind injection wasn't there would it be biased the other way—there would be a lot more hysteria, a lot more pressure to throw it out because we're worried, you know would it get airplaned.

A: You know, we have this problem, when do we excite the LSC and when do we excite the entire world? I think to excite the LSC [Detection Committee] a one in ten-year false alarm rate is good enough for me. . . . And for the external world, its one in a hundred.

C: The original number I heard for a primary search . . . was one in one thousand days.

4 The Equinox Event: The Middle Period

As October turned into November 2007, the talk was of how fast to move the analysis forward. The first question was whether the event should be written up and passed on to review committees before the box had been opened on the main analysis. After a lot of discussion it was decided that this could not be done—the full analysis had to be completed first, and this would take time.

The argument was that only after the full analysis had been done could one be sure how unlikely the Equinox Event really was. Only after the full analysis had been done would one really understand the background. If the background were "stationary"—that is, even across the whole stretch of data—one could do as many time slides as one liked on a short stretch and develop as much understanding of the background as one liked. But if the background is nonstationary—that is, some stretches are more heavily populated with glitches than others—then a longer stretch of data is needed to handle the background with more confidence. This means that the effective sensitivity of the detector becomes a function of how long it is switched on. Suppose one had switched on the machine for only two months and seen such an event. Would one have to say: "We cannot count that as an event because the machine was not on for long enough to reassure us that the event was unusual"? Or, to look at it another way, should S5 had been extended so that the extra time on air effectively rendered the machine, not only more likely to see something that might occur, but also innately more sensitive because one could speak more authoritatively about the unlikelihood of anything that did turn up. So it turns out that sensitivity for even a single event is a function of time on air, and this is not something that seems to have been written into the design of the apparatus. Up until this point the logic always seemed to be that likelihood of seeing an unusual event, but not sensitivity, was a function of time on air. Another very good reason for staying on air longer is that seeing a second event would markedly change the confidence in the observations.

In any case, it was hoped that the full analysis would be completed by January or February 2008 and that, therefore, the envelope could be opened in the collaboration meeting of March 2008. Actually, it was not to be opened until a year later—a full eighteen months after the equinox of 2007—and even then, for some, it was opened prematurely.

December 2007

There was another meeting in the second week of December in Boston, where things seemed to be dragging more and more. Below are some extracts from my notes made at the time. In reading these notes it should borne in mind that the sociologist's interests are not always the same as the scientists' interests, though a good few of the scientists share some of the sentiments expressed below. A clean and unmistakable first discovery would be a sociological anticlimax just to the extent that it would be a glorious scientific triumph; the sociologist would have much less to talk about. Likewise, insofar as upper limits are scientific successes, they are of little interest sociologically except that they are perceived by scientists as being a success. Results that the scientists have to argue about are much better from a sociological point of view, for they display the process of decision-making. Here, then, is an account of what was happening around this time from the sociological perspective:

> I sit through paper after paper that finds something initially interesting but winds up showing that it can't be separated from noise. . . . All this "can't be this, can't be that, must be less than this, must be less than that,"—it is all beginning to get exhausting and numbing. Why is there no trace of gravitational waves *anywhere*. Oh for the days of Joe Weber and the Italian claims.
>
> [Someone remarked] that even if, after they open the envelope, they find the Equinox Event was not an injection, they still have to check to make sure that no-one injected anything maliciously and to do this check properly will take a very long time. One could go on like this forever—there is always

another check to do and then another you can dream up; the structure of checking is just like the problem of Tantalus—there is always some new check you can dream up. And this is where things are going. *The apparatus is becoming less and less sensitive*. The equinox event satisfies all the criteria set out before the analysis but now it has been decided that it is not enough. A new criterion has been invented—that the individual contributors to a coincidence must not look too much like typical bits of noise. This means the apparatus has another blank spot in its detection footprint. The scientists say it can see "this far" but now it is working they say "it can't see this far after all." . . .

There is no excitement about this event any longer as there was in Hannover. . . . Is it that the injections have taken the fire out of every ones' bellies? Is it the weight of history pulling everyone down? In a casual remark to [. . .] I point out that I know of no episode in the history of science where the scientists were commended for handling history properly—science is discovery, not non-discovery. Is it just fear of saying something unclear—because the equinox event is quintessentially liminal and cannot be handled? People are scared of having to say what they will have to say—"they are not sure." But as even [. . .] said, the most likely event is a liminal event. And this means if you eliminate liminal events you did not tell the truth about the sensitivity of the apparatus.

The Start of the Middle

What was also going on in this middle time was further refinement of the analysis. Various things were happening. More time slides were being carried out, so that the background became better defined. Work on additional pipelines was being completed, and one of these took into account, not only the fact of the coincidence, but also the coherence of the signal on the separated detectors—did it make sense that the signals seen on the detectors could have come from a single point in the sky representing a single source? The answer from this pipeline was "yes," and that implied a major increase in the likelihood of the event being real, or the unlikelihood that it was a chance coincidence.

The following figures can be read as a visual illustration of the kind of statistical analysis that was going on. These figures were developed after the position and frequency of the putative signal had been extracted by the computer; knowing these features of the signal, the appropriate filters could be applied to the raw data to make it visible in the data stream. The figures, then, do not represent the process of extracting the signal statistically but may help in getting a sense of what kinds of judgments were implicit in the statistical analysis.

Across the whole frequency range, noise swamps the raw output from the interferometers. Low frequency noise completely dominates any putative signal, and when examined the raw data looks like a smooth curve. It is only after these very loud noises have been removed that any structure begins to emerge. Fortunately, the sources of the large noises are sufficiently well understood for them to be filtered out without risk of distorting any gravitational wave signal that is there; the whole art of interferometer design is to allow noise at frequencies well away from gravitational wave signals but to minimize noises that occur within the gravitational radiation waveband that the machine is exploring.

Figure 4 shows the result of the application of stages 2, 3, and 4 of a filtering process, informed by what had already been done in the statistical analysis, to the run of data from an interferometer which includes the Equinox Event. The first trace shows what is left after the very loud low frequency noise is removed. The result is a trace dominated by loud high frequency noise. Removing that leaves the middle trace. Here the shape of the Equinox Event begins to emerge, but it is still confounded by other noises that now do fall within the waveband of interest. Fortunately, some of these are also well understood—for example, the noises caused by vibration in the mirror-suspension wires are well understood and well defined—so they can be eliminated by narrow filters. Examples of such imperfections can be seen in figure 3. They are the tall spikes at certain narrow frequencies. After the last filters eliminating these spikes are applied, the signal stands out fairly clearly, as can be seen in the rightmost trace.

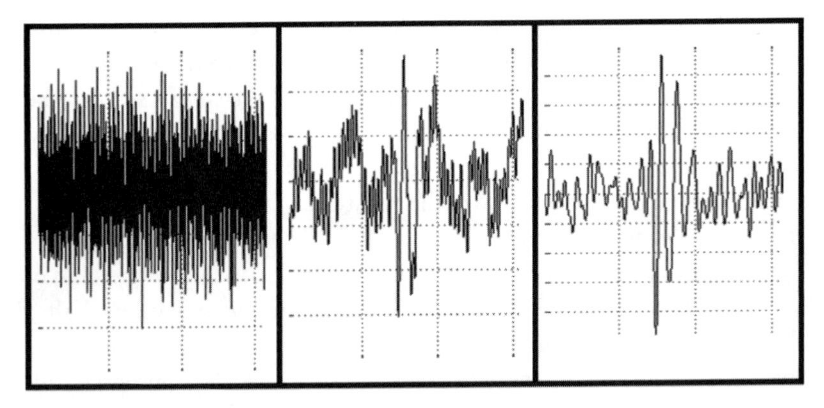

Figure 4. Stages 2, 3 and 4 of a filtering process that reveals the initially hidden Equinox Event signal. (Traces by Jessica McIver.)

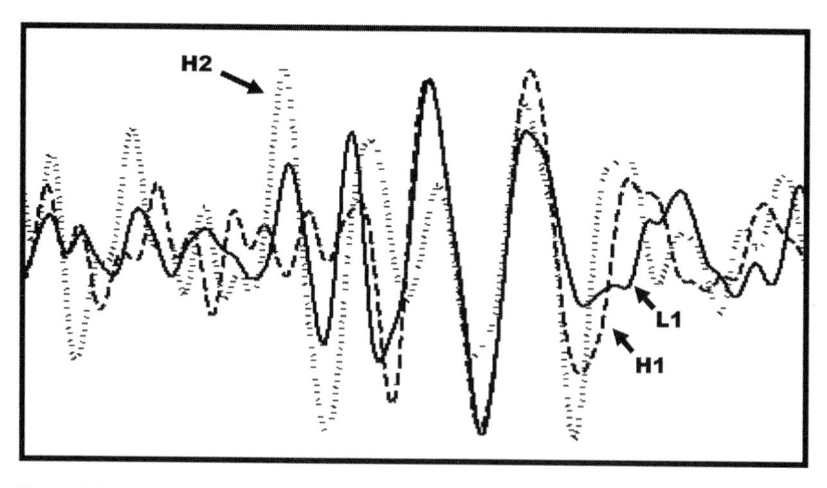

Figure 5. Equinox Event: H1 (dashes), L1 (solid), H2 (dots). Jessica McIver did the scientific work to produce the plots shown in figure 5—designed to be viewed in color—for the LSC-Virgo collaboration. She was kind enough to make new versions suited for black-and-white reproduction especially for this book.

Figure 5 shows enlarged versions of the filtered Equinox signal with the outputs of H1, L1, and H2 overlaid. The statistical analysis which is being discussed in the telecons reflects, in effect, the extent to which lines like these fit over each other, though the lines are, as it were, virtual—they are something like what is implicit in the statistics. The confidence expressed is a measure of the rarity of such close-fitting high amplitude outputs when it comes to overlaying the noise on noise as revealed by the time slides exercise.

To put the matter another way, one has to look at the extent to which coincidentally high amplitude signals from the three detectors, which we can think of as being represented by something like figure 5, overlap. One then has to ask whether this signifies the effect of some common external cause on the three detectors. The only way to know is to see how often the signals might overlap in this way if there were no external cause. We know there is no external cause when time-offset outputs are being compared. The answer obtained from the time slides is that something like this comes up about twenty times in six hundred years, or roughly once every twenty-five years. Given no other sources of information, that is what a discovery is! It is the decision that this event is so unlikely to have happened by chance that it must represent something "real." Hence it is easy to understand much of what will be found in the remaining chapters. We will examine the role of what is culturally acceptable in the scientific community as a publication and what counts as an acceptable discovery as opposed to an outrageous claim; each of these can change. We will look at a carefully worked out flowchart for the process of decision-making. And we will attend to arguments about the exact wording of the way any putative finding or non-finding is to be announced to the world.

More work was also being done on vetoing out noise so that the background would go down. If the source of a noise was understood, it ceased to be merely a random noise and became instead a disturbance that could be eliminated from the analysis, reducing the number of chance coincidences. (This is a dangerous game to carry out post hoc because one might be tempted—subconsciously—to remove noises that do not look like the "event" while not removing those that do look like the event.)

By the telecon of 19 December 2007, confidence was growing that the Equinox Event would jump the hurdles set by the group and be taken on to another stage. The following extract from that telecon gives a good sense of the state of play and the difficulty of turning a blip in the statistics into the binary "yes" or "no" of a "discovery."

A: [Does taking this] to the Detection Committee mean that the Burst Group thinks this is a gravitational wave? Is that the take that other people have? . . .

B: A, you're probably right, but the definition of what is a detection candidate is subjective, and I don't think it can be unanimous. That's why I think we should be presenting it, not necessarily as our detection candidate—we can present it as one event in ten years. We should present it in mathematical terms rather than put in words what this is about. And we should ask the Detection Committee to view it like that. . . . We should present exactly what is the significance of this event and we should ask them to view it with that in mind. . . . For someone a detection candidate might be at the level of one every hundred years, for others, one every ten years, and I don't think it is our goal right now to agree exactly what that means within the group. As long as we agree that outstanding events should be brought to the Detection Committee for further consideration by the collaborations. . . . [We have heard] when in doubt, we should bring things to the attention of the collaborations.

C: [But] we cannot just present something to the Detection Committee, saying . . . we give this for your review and if you like it we will publish it and if you don't like it we will not publish it. It's not the way to go. The Group should stand behind it and have a clear opinion about it. . . . The Group should defend its opinion.

B: But if you remember [the] flowchart, at the end of the day there is going to be a vote of the Council, in order to decide whether we publish or not. . . . We can have an internal vote based on what we have in hand right now, and if this is the only event we see in the S5 [and Virgo] run, and if that statistical statement on the significance of the event is correct, is that crossing everyone's threshold for publication or not?

D [This is a paraphrase]: We must have a consensus, but the consensus might be to make a weak statement about it. . . .

B: . . . I will be delighted if the Burst Group really has consensus on what exactly this event is and whether it should be published or not. But I would not be surprised if there are opinions all over the spectrum from "this is background we simply don't understand" to "this is probably interesting but we cannot publish" to "go on and publish." So I think we should be prepared within the group and within the collaboration, if it goes down this road, to have opinions that are going to fall over the spectrum. And I think it is fair given that most of the work we have in hand right now does not have this event as maybe the gold-plated event that one would like to see for the first detection.

Glitches

At a meeting in December 2007, one of the Burst Group members had remarked that their confidence in the event was growing. But another member responded: "But our confidence in our confidence is not very high." What he was getting at was something else that was pulling in the other direction. Right at the beginning of October, when I had telephoned around to gauge the level of interest in the event, I had found that one of the Burst Group members had said he was not very excited because the event was right at 100 Hz, which is where all the glitches were found. In Hannover I had told another respondent that, as a sociologist, I was depressed by these glitches, since it seemed to make it unlikely that the event would ever amount to anything. The more it amounted to, the happier I would be, because the more sociology I could get out of it—not to mention my "he's finally gone native" sheer excitement about the possibility that a gravitational wave had been discovered. This respondent had tried to reassure me, however:

Respondent: Well, some days it depresses me, and sometimes I say "well, that's why we have statistics." Because you get to ask yourself the question—this event happens so often in H1 and so often in L1 and how often do they happen at the same time with roughly the same degree of match as what we see here? So it is in fact possible in principle to overcome that. But it has to be something about the characteristic of the event; it is extremely unusual even if its building blocks are usual. Like, for example, the individual events are seldom this strong—we know they were simultaneously strong in both. And they happened so close in time that the wiggles really overlap. . . . But we've known the data had glitches, yet we look for things that shape. We've always said we could find . . .

Collins: In other words, the shape of the glitches is a reasonable shape for a signal as well.

Respondent: Yes they are—they are! Absolutely. We're not being lunatics, we're just doing something that's dangerous. . . . And that is the depressing thing that I learned this summer [during a project in which I was trying to separate glitches from event candidates. Glitches look like signals.] . . . [E]very single glitch in this detector has only one [signal like] pattern [by and large].

Now, however, the "it's too much like a glitch to count as anything" argument was growing in momentum at the same time as the statistical significance was growing. On 11 January the same respondent emailed me with the following remarks on the matter, referring back to a telecon of a few days earlier:

> H1 shows a number of other glitches within 10 sec of the equinox event. . . . There was no agreement on the ultimate question of whether background rates are thus mis-estimated. . . . People are going to do more thinking about this, and there will be more discussion in the future about the extent to which one should say that H1 was stationary or else if it was misbehaving at the time of the event. (I'm in the "misbehaving" camp now.)

The detectors all have systems for vetoing data which are not of good quality. Where the data is bad, a "data quality flag" is set, and the data are thrown away or used only with great caution. No data quality flags were showing in the areas of data contributing to the Equinox Event, but it was now being argued that this was a problem of the procedures. Close examination of H1 showed a lot of glitches near the time of the event. These had not been noted because the time intervals within which vetoes were set were too coarse to pick out this area as any different from other areas. More refined examination was showing that the contribution made by H1 to the Equinox Event looked just like a number of other glitches that were happening in H1 around the same time. If the contribution of H1 was really a glitch, and it was looking more and more as though it was, then what was being seen was not a coincidence between random noises, nor a demonstration of the effect of a signal, but a correlation between a noise in L1 and an artifact, that was understood and expected, in H1—in other words, something of no interest at all. Let us call this "the glitch hypothesis."

The First Positive Account

In spite of these doubts, by August a junior member of the team, whom I will call Antelope, had written up the results in a report which spoke of a "Burst Candidate." The existence of this report became widely known and gave rise to a heated telecon on August 20th. I found out about it from emails and minutes. What follows below are extracts from the minutes of a telecon, with a brief introduction to it by a respondent. They give some sense of how the scientists think about a possible detection:

> There was a heated telecon when it became known that Antelope had started writing up a Detection Report on what he is calling "Burst Candidate 070922a" which takes a rather enthusiastic approach to the E.E. People wanted to know why the Burst Group was sitting on this possible detection, instead of immediately forwarding it to the Detection Committee. The main points of the discussion as recorded in the minutes were as follows:
>
> A: need to update 1 per 50 yr number based on Year 2 analysis, when we open box in few weeks . . .
> B: there will be a total of 1000 time lags in several sets.
> D: don't you know already that this is of least 1 per 10 years?
> B: you know from many ways it is even stronger
> E: even 1 per 5 years is interesting
> B: we don't have good statistical measure just from FAR [False Alarm Rate], look at probability distribution—this is less than two sigma. . . . In most experiments, don't look at 2 sigma results.
> D: Need to move forward with process, this is strong enough
> A: We might decide that at 2 sigma we can't claim a detection. . . .
> G: The system has failed. It has taken too long to finish this. Statistical perfection isn't the best goal. . . .
> D: Send this now to Detection Committee?
> H: This would be great

I: Don't think this should be a claim presented to a jury
E: Will never be unanimous
J: There are other purposes to bringing this forward. For ex: Does this influence commissioning and running?
H: We know there is a problem with glitches that look like this event.
I: agree that this a big problem
D, J: we need to know whether it is an event or a glitch
H: This is hard to do, given the population of glitches
F: Are we procrastinating?
E: No, Burst Group isn't procrastinating. . . .
G: We need to find a better way to engage the Collaboration.

The Amsterdam Meeting

In September 2008 the collaboration came together again in Amsterdam. The group was trying to work its way to a consensus that would be acceptable to the wider collaboration on how the Equinox Event was to be

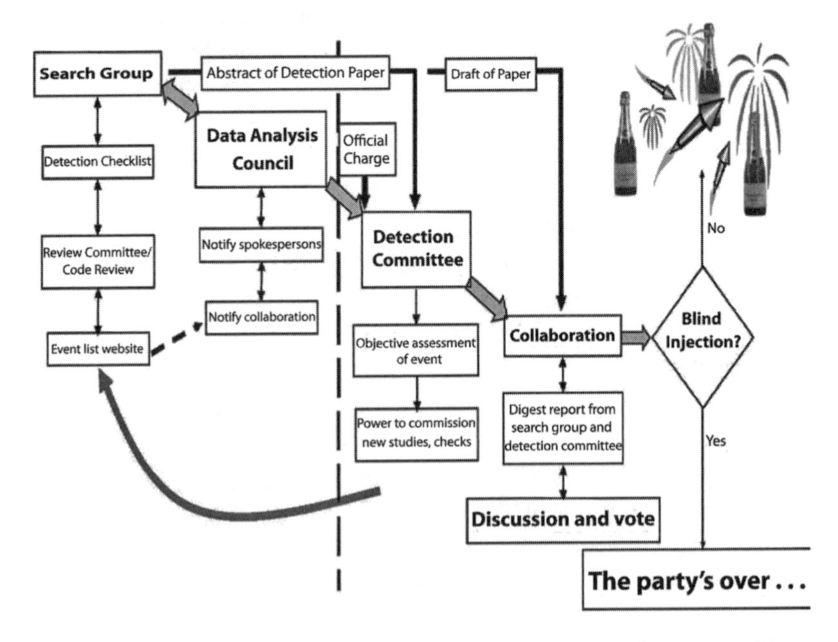

Figure 6. Organizational chart for making a discovery in LSC-Virgo. This version of the chart was first presented at the Hannover meeting.

handled and spoken about. One thing I have discovered about physicists, or at least this group of physicists, is that they love to try to solve problems by inventing organizational structures. I have often been surprised that, when I have asked a question of a senior member of the collaboration about what the members are thinking about this or that conundrum of analysis or judgment, the reply refers to the committees or bureaucratic units they are putting together to deal with it. It is as though a properly designed organization can serve the same purpose as a properly designed experiment—to produce a correct answer. At the Amsterdam meeting the collaboration was reminded of the institutional process that would lead to a discovery by means of the organizational chart shown in figure 6.

In case readers have any lingering doubts about the social nature of scientific discovery, this PowerPoint slide should dispel them. It is a matter of social organization. Furthermore, as can be seen, the "proximate cause" of the first discovery of a gravitational wave will be, not a physical event, but a vote. As was explained in the context of the "airplane event," a vote is a sure sign that something sociological is going on.

The Abstract

There were two talks about the Equinox Event in Amsterdam. The first was a description by the group of what they thought they would say about it in the proto-publication that was being drafted. This description would take the form of an "abstract" to be presented to the meeting. Days were spent arguing about what this abstract should contain. I collected several drafts as the final version evolved. An early version contained the following wording:

> The estimated rate of the background events with the same strength or stronger is once in 26 years. . . . From examining publicly available data, no electro-magnetic counterpart has been found. The observed sky location is not consistent with the galactic center or the Virgo cluster. . . . Because of the moderate significance of the event, and its close resemblance in morphology to the expected background, we claim no detection.

A subsequent version read:

> A single event above these pre-determined thresholds resulted. It is of moderate significance, XYZsigma, and of resemblance in frequency and

Figure 7. Members of the Burst Group draft the abstract to be presented to the Amsterdam LSC-Virgo collaboration meeting.

> morphology to background events. On this basis, the event is not considered as a genuine gravitational wave burst candidate.

The next and final version, largely developed during the lunchtime discussion shown in figure 7, but with some subsequent modifications, said only:

> The analysis yielded three events, which were above the search thresholds based on measured background rates. One of these events passed all veto conditions which we had established in advance. We examined the events in detail and we do not consider any of them to be genuine GW candidates.

Figure 8 shows the moment in the presentation when that sentiment was expressed as a PowerPoint slide. The last slide in this talk (figure 9) is a set of bullet points summarizing what had been said.

The Equinox Event was not going to expire that quietly, however. One senior member of the collaboration clearly found the negative conclusion galling and asked how they could be so sure it was not a gravitational wave candidate.

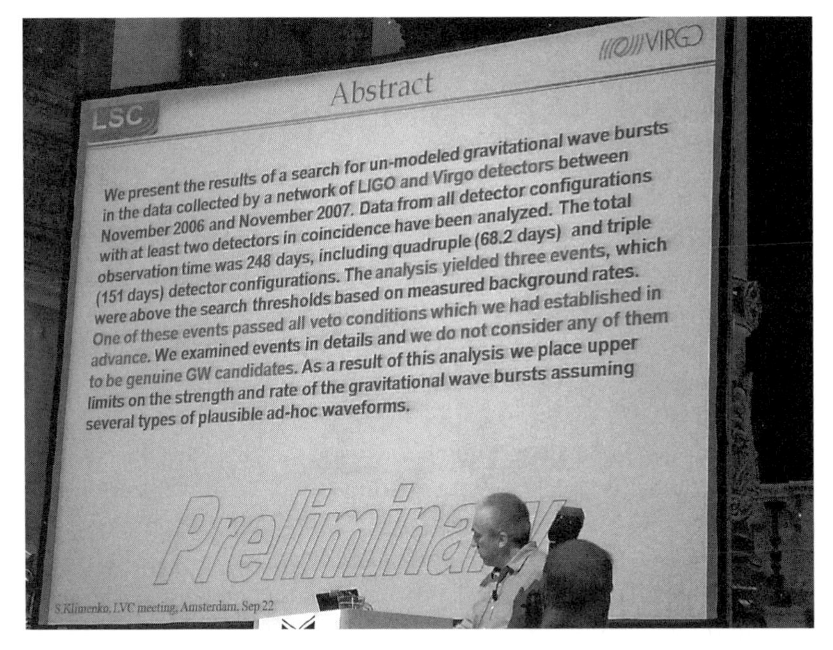

Figure 8. In Amsterdam, the Equinox Event is announced not to be a gravitational wave.

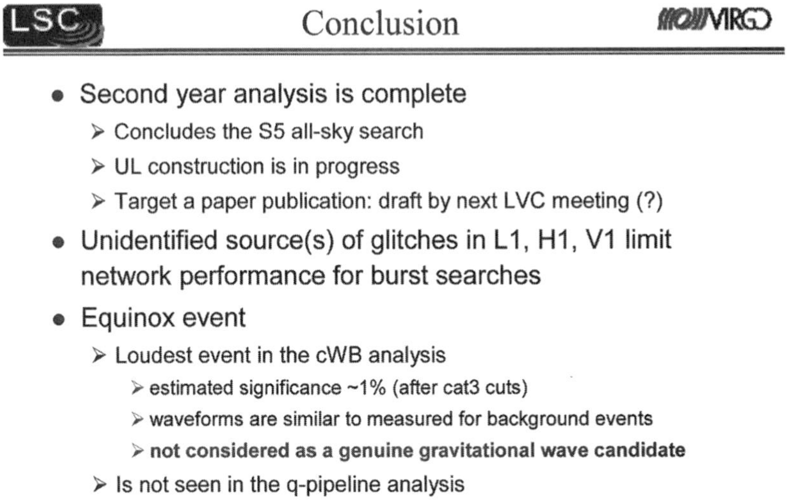

Conclusion

- Second year analysis is complete
  - Concludes the S5 all-sky search
  - UL construction is in progress
  - Target a paper publication: draft by next LVC meeting (?)
- Unidentified source(s) of glitches in L1, H1, V1 limit network performance for burst searches
- Equinox event
  - Loudest event in the cWB analysis
    - estimated significance ~1% (after cat3 cuts)
    - waveforms are similar to measured for background events
    - **not considered as a genuine gravitational wave candidate**
  - Is not seen in the q-pipeline analysis

Figure 9. Equinox Event riddled with bullets

A second senior member made a point against the glitch hypothesis. The second bullet point in the last set of three expresses the glitch hypothesis: that, since the event looked like a glitch and occurred in a part of the spectrum full of glitches, at least one of its components was likely to be a glitch. He said:

> I don't understand why the second bullet is a significant consideration: That tells me only that whatever astrophysical event this may have been was unresolved by our instrument with 100Hz bandwidth. . . .
>
> Our instrument has a finite bandwidth. If you can't resolve the event, everything will look like the frequency response of the detector.

This turned out to be a key argument against the glitch hypothesis. The detectors have what is called a "response function." Though they are said to be broadband instruments that can follow the intricate ups-and-downs of a gravity wave as it passes through them and draw them on its output, once more nothing is so simple, and the output is a function both of the waveform and of the resistances and affordances of the elements of the complex machine itself.[1] The second senior member was saying that the machine cannot always "resolve"—that is, exactly follow—the waveform of events because its responses are not equally fine in different wavebands, or areas of the spectrum. Therefore, he said, what hits the instrument might very well look like a glitch, so the fact that this event looks like a glitch does not show it was noise rather than a gravitational wave even if it looks like noise rather than a gravitational wave.

Here is the tension between statistics and craftsmanship that will not go away. On the one hand, the premise upon which calculations of the sensitivity of the detectors have always been made is that it is all a matter of statistics. One the other hand, the upholders of the glitch hypothesis feel themselves empowered to say that the meaning of statistics can be adjusted as a result of a craftsmanlike examination of the behavior of the device at the time the data were recorded. As one of my respondents put it in an email:

> This is what you call "double counting" [see below], and what [has been condemned as] "seeking reasons not to believe," but I think that it is a highly responsible use of "experimentalist's craft judgment," which must . . . be applied especially for the very attention-getting case of a first detection.

1. "Affordance" is a philosophical term: e.g., a door handle is designed to "afford" turning.

One danger is that the outcome of such an examination cannot be legitimately used to enhance the statistics—that would be post hoc data massage—but it can be legitimately used to dilute the statistics because that is a conservative move. A conservative mindset, then, has opportunity to exercise itself.

Antelope Again

I have to confess that I do not understand the organizing principle behind the agenda for these meetings, but it is nice for this account that, after all this, Antelope's paper told a different story. His summarizing, bulleted, PowerPoint slide is figure 10.

As can be seen, the first entry under "The Interesting" reports an independent analysis by a scientist we will call "Bison" and his group. This had come in too late for detailed evaluation. The Bison group's approach produced a likelihood of only once every three hundred years for an Equinox-like event to be produced by noise alone. If the Bison group's approach were taken seriously, the previous presentation's report of the Equinox Event's death would certainly have been premature. The majority of Burst Group analysts, however, felt that the Bison group's results were unreliable

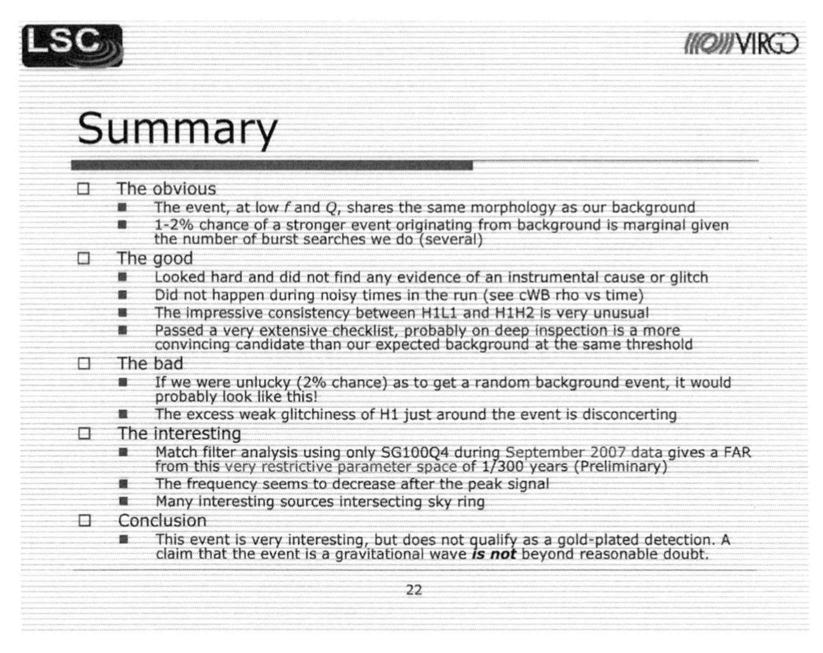

Figure 10. Rumors of the Equinox Event's death may have been greatly exaggerated

in consequence of the "trials factor" (see chapter 5)—the Bison group had tried too many ways of analyzing the data before reaching a conclusion.

The final bullet also runs against the tone of the previous presentation, which was represented as approaching group consensus. Instead of:

> [The Equinox Event] is not considered as a genuine gravitational wave candidate.

We have:

> A claim that the event is a gravitational wave *is not* beyond reasonable doubt.

Note how much effort went into the drafting of the exact wording of the abstract presented by the Burst Group and note how much difference the slightly varied wording in Antelope's version makes. This is a theme that will recur. On the one hand, we have physical science—that quintessential home of the supposedly exactly calculable; on the other, we have hours of argument over the exact wording to be used in describing a passage of scientific activity.

Political Interests

The gravitational wave collaboration that comprises the LSC and Virgo has now gathered into itself nearly every person in the world who has any expertise in gravitational wave detection physics, and if science can only remain healthy so long as what Robert Merton called "organized skepticism" is encouraged, then that skepticism had better come from inside the collaboration. Barry Barish, who led LIGO to the point of reaching its design sensitivity, frequently made the comment that the only skilled critics of the work being done were to be found inside the collaboration and that, therefore, the organization had to work out its own internal means of subjecting its work to critical scrutiny. The discovery flowchart (figure 6) institutionalizes this principle. But the collaboration can also be seen as a series of healthily competing analysis groups, each ready to try to dismiss each other's findings.

Thus there seemed to be some competition between members of the Burst Group and members of the Inspiral Group. Members of the Inspiral Group seemed more prone to be dismissive of the Burst Group's detection candidate than were members of the Burst Group itself. A senior theoretician who contributed much to the methods of the Inspiral Group made a statement—and it is a statement that he repeated at most of the meetings I was go to—to the effect that the Burst Group on its own could never make a detection. The underlying argument was that a signal profile from the Burst Group could fit no known template—if one did, it would be another group's business—and that unless such a formless signal turned out to correlate with some other event visible in the electromagnetic spectrum, it could never be separated from noise. He said of the Equinox Event and by implication of other similar events that might be found by the Burst Group:

> We don't know where it is; we don't know what it is; and we're not even really sure that we saw something. It's hard to publish that. That's the difference between these searches and the other ones. Right?

Another leading member of the Inspiral Group referred to the conclusion of Antelope's paper and said of it:

> I would never put that statement in the paper because we simply do not have enough information to say that, so we simply couldn't make that claim. The fact is that the truth really is something that we probably will never know [unless it's a blind injection]. We'll never have access to events that are at this level of SNR, so the claim that the event is a gravitational wave "is not beyond reasonable doubt"—I think the claim itself is [meaningless]. . . . Whether the event is a gravitational wave or not we may not be able to answer.

Once more, this remark, if taken seriously, would make it very difficult for the Burst Group ever to claim they have found anything unless it were so loud or so well correlated with other events that the group's refined statistical procedures became almost irrelevant.

Another disagreement over the value of the Equinox Event was to emerge between LSC and Virgo. The LSC wanted to use the blind injection challenge in general, and the Equinox Event in particular, to rehearse the entire detection process. The consensus seemed to be that any event that was likely to occur by chance in the region of less than once very ten years should proceed through the detection flow chart right up the Detection Committee—the last stage before it was taken back to the collaboration for a vote. It should be completely clear, by the way, that not a single person I ever spoke to, nor me, for what it is worth, believed that the Equinox Event ever had a chance of gaining the imprimatur from the Detection Committee of "worthy of publication as a discovery claim." The following passage of debate indicates the tenor of discussion at the Amsterdam meeting:

> A: . . . We should actually be able to state how many sigma the candidate is. Gosh, if we can't state that we're in real trouble. [Interjection—we can state it for this one—it is 2.5 sigma—laughter.] You tell me it's 2.5 sigma, […] says he doesn't believe it, other people here say we should use trials factors [see chapter 5], OK! I would like to feel we could arrive at a consensus about what it is and the collaboration stands behind the . . . [inaudible] sigma, and right now I don't hear that.

From the floor a question is put about what would count as a detection claim.

> A: I don't have an answer to that, and it goes back to what I said yesterday. What we did in the Inspiral Group and we asked the question—"What's interesting?" And its, kind of, one per one hundred years. But that's the low end of interest. And we aren't sure if it's one per thousand, or one per ten thousand years, that is really where we start to feel comfortable. We don't know the answer to that. So, honestly, as far as a criterion for stating, "we have a detection," my impression is that the Inspiral Group feels that we have a two orders of magnitude uncertainty in the false alarm probability.
>
> B: . . . [P]eople do this analysis in high energy physics, and they find 3 sigma all the time and they usually disappear when they do the analysis again. So 3 sigma is not significant at all. 4 becomes interesting, and 5, "this is serious."
>
> C: You can't define sigma and the reason is the long tail. [Statistical analysis is based on models which follow smooth patterns. Here the pattern is not smooth because of the glitches—which show up as an anomalously long tail.] . . . If the long tail persists and then you've got to argue about whether . . . you are willing to say "this long tail kills us" and "we just don't know how to get rid of the long tail," then you've got to deliver it. On the other hand, if you are willing to go out on a limb a little bit, and say "hey, it's a long tail, it's very improbable, one or two percent is, you know, one in a hundred years is definitely of interest, because you've got the tail"—you would never say that if you didn't have the tail—then yes, it's a different problem.
>
> D: I cannot imagine us making a first detection just based on some number—you know, one per one hundred years or something—may well go

above that—"that's it, we've made a detection." If it's one per ninety-nine years then negligible. So . . . we'll have some reasonable number like one per one hundred years, and then when it comes close we'll follow up and we're going to claim the first detection because we also saw that event in a telescope or in a neutrino detector as well, or something like that.

The Equinox Event, as can be gauged from this discussion, just did not have the "gravity" to be the golden event that would convince the world that a first detection had been made. It was all a matter of nuance—Antelope's account versus the accounts that were more hostile to it being discussed as a potential, though regrettably weak, real event. Nevertheless, the Event did seem to meet the LSC's criteria for going before the Detection Committee for further examination. In spite of this, members of the Virgo group were adamant that it should not go even that far. LSC people, on the other hand, wanted to exercise as much of the detection procedure as was possible and were less than happy that the procedure had been cut short.

Why did the Virgo group take this negative attitude? Once more I will invoke the Anti-Forensic Principle (see the start of chapter 3): I don't know! What I do know is that two competing accounts were in circulation. One was that Virgo did not want to see the Equinox Event treated too seriously because Virgo was largely a bystander (except insofar as they shared in the data analysis). Virgo was not included in the blind injection challenge because the protocols were not ready. (Remember, the Equinox Event fell into a waveband where Virgo was much less sensitive than LIGO and thus could not be seen by Virgo; this was something that encouraged analysts to guess that it was a blind injection and put less effort into analyzing it.) This account, then, said that Virgo did not want to see significant gravitational wave-related activity going on unless they were involved.

The other account, given to me by a senior Virgo spokesperson (S), was that Virgo was more sensitive to the way these things can get out of control than the LSC because they had seen it happen in their own backyard with the Rome Group. They understood the Detection Committee to be there to give the imprimatur for publication, and they wanted to cut off any such possibility from the outset. After agreeing that Virgo had indeed "vetoed" the promotion of the Equinox Event to the Detection Committee level, he said:

S: I think that much of this has to do with the past history of several crucial Virgo members who were operating bars and so they had previous experience of possible announcements that then went into nothing. And so this makes us extremely cautious because what has happened is the way this Detection Committee was formed, which is in disagreement with what [I later heard was the purpose of the Committee and this was that the Committee] was really the ultimate judgment before publication. And so people felt that the Equinox Event was not at the level to go to that step—this is the reason. I think there has been a different perception of the role of the Detection Committee and the other point is this past history of people.

Collins: So you think that the American version of the Detection Committee is really more something that will rehearse and is quite likely to throw things out?

S: Yes—I talked a lot with [. . .] and he was saying that he was seeing the Detection Committee as something that would try to investigate further and even, in a sense, sweep away the fears from the analysis group by asking other questions. Also one thing that was one of the worries of [. . .] was that the analysis groups don't have too much of a grip on the detector itself. They analyze data, and they have less experience of really having the detector working and so knowing that it's a delicate instrument. And so I think this was one of the reasons for the Detection Committee as [. . .] was putting it. But then, it was presented in a drawing, where there was a flow diagram, as the ultimate judgment, and so . . . and we were afraid of collective madness.

C: Some Americans think there is another problem—which is that Virgo does not want anything to be a detection until Virgo is sensitive enough to have a positive share in it.

S: I think this is not right. . . . Absolutely not.

C: Honestly, hasn't anyone from your group ever pressed this on you?—Don't let's let anything go forward until we've got Virgo running?

S: No—no, absolutely not.

C: So if it turned out that this Equinox Event had been bigger, you would have been happy to see it go forward?

S: Yes.

S then pointed out, quite correctly, that there has to be a possibility that one detector team will see an event and other will not because, depending on the characteristics of the source and because of the different orientations in three-dimensional space of the LIGO and Virgo detectors, the signals seen in one may not be seen in the other.

On the other hand, if one wanted to make a case for the more political interpretation of Virgo's resistance to the highlighting of the Equinox Event, one could point out that they also resisted, insofar as they could, any positive mention of it in the draft abstract for the publication to be discussed below, where, at best, it would have been said to be a candidate that had insufficient statistical significance to be counted as a discovery, so that no "collective madness" could have ensued. Of course, Virgo, like the LSC, is a big group, and not everyone in it may have had the same motivations.[2]

Whatever their motives, the Virgo group stopped the Equinox Event going forward to the Detection Committee. Many LSC members believed that, as a result, a valuable opportunity for rehearsing procedures, which, if stringent enough, would give the analysis groups more confidence to forward insecure claims, had been lost.

In terms of the metaphor offered at the outset, the net outcome of all these debates was that Equinox Event was never allowed to become a lasting eddy in the stream of history—nothing within which either new physics or new social life could reorganize itself in an enduring way. The Equinox Event had stilled the stream for a moment, but now it was lost once more in the turbulent water of the social and physical world.

2. On the other hand, in later private correspondence (October 2009), S said that he saw the problem being not so much with insiders as with outsiders taking any result, however poor the statistics, and trying to find some event in the heavens that correlated with it. This could give rise to a lot of speculation and "noise" in the press that had nothing to do with real gravitational waves.

5 The Hidden Histories of Statistical Tests

In today's frontier astronomy, all one "sees" are numbers. The numbers represent the things that would once have been seen in classical physics (though these may subsequently have been counted and represented as numbers). The great advantage of looking at numbers rather than things is that it makes it possible to impute the existence of things that would be too faint to spot in any more direct way. Thus, if you want to see some faint object in the heavens, you have to allow its emissions—light, radio waves, X-rays, neutrinos, or whatever—to have an effect on you. If you use your eyes as the receiver, you need a lot of emissions. But if you use an array of fancy electronics that can count the impact of individual photons over a long time, you can calculate whether there are a few more photons coming from this point in the sky rather than the background, and that can tell you that there is something there, emitting photons, but at a much lower level than would be required to make much of an impact on your retina. So the development of cleverer and cleverer ways of gathering fewer and fewer photons to make an ephemeral "spot" means that the frontier of observation is always moving out and taking fainter and fainter objects into our observational purview.

The cost of this progress is having to move straight to statistical assessments of the significance of numbers without the prior intervention of the eye.[1] Of course, there is no such thing as a

1. Sometimes misleadingly represented in the newspapers with false-color computer images that look like photographs.

"direct" observation, but sometimes the degree of indirectness seems to stretch the very idea of "observation" to the breaking point.

If it weren't that seeing is done today with numbers, there would be no gravitational wave detection science, because gravitational waves are far too faint to "see" in any other way. The science is a matter of the most intricate calculations meant to extract meaning from tiny electrical currents generated within each giant interferometer, as these currents struggle to compensate for—and thereby hold the whole apparatus in equilibrium—mirror-movements thousands of times smaller than the nucleus of an atom that are bought about by minuscule changes of the average phase of photons circulating within the interferometer.

Going back to photons emitted from some point in space, the question is whether there are "a few more photons than there should be" coming from "that" point as opposed to the background, or that might be caused by light scattered off dust in the atmosphere, or whatever. Inevitably, the calculation involves statistics: "There are a few more photons coming from that point—but could it be just a matter of chance scattering that has created a bright spot, or does it mean there is really some object doing some genuine extra emitting?" In the case of gravitational waves: "Is this movement of the mirrors, represented by those numbers, something special, or is it just the kind of random movement that is going to happen every now and again in any case?"

As has been seen, the answer comes in the form of "unlikelihood." An "observation" consists of a statement: "It is so unlikely that this concentration of photons/set of coincident movements of mirrors could have arisen by chance that it must represent something real."

In the published papers and the announcements that win their authors Nobel Prizes, that unlikelihood is represented by another number. A number is the kind of thing we think of as "objective." That is to say, numbers appear to arise as a result of well-defined states of affairs in the world. It is possible to argue about whether this is a big pile of apples or a small pile of apples, but if there are fifty-six apples, then "that's it." That is why decision-makers love numbers—the apparent "that's it" quality of numbers seems to relieve decision-makers of the responsibility of judgment.

Frequentist and Bayesian Statistics

There is a long-running argument among those who use statistics in the physical sciences between the Bayesians and the Frequentists. Indeed, the argument is of such long standing, and gives rise to such passions, that people on either side sometimes jokingly refer to their preferred preference as their statistical "religion." It is worth looking at the religions for a couple of reasons. First, examination of the argument between the religions is one way to begin to show that the interpretation of statistics is always subjective, however objective it looks. Second, the Bayesian approach, as I shall argue later in the chapter, can be used to justify a publication strategy involving weak claims. This chapter starts with the first reason and ends with the last, other elements of statistical subjectivity being explored in the middle part.

The crucial difference between the two statistical religions seems to be that the Bayesians believe that the unlikelihood statement that is at the center of all statistics-based claims must take into account what you already believe about the world—the "prior probability" of the claim being true. The Frequentists believe that prior probabilities are too subjective to feature in a statistical calculation which results in a number and that the number, therefore, should reflect only what likelihood you calculate, not whether that calculation is credible or not.

Now it is obvious that prior credibility plays a part in the assessment of unlikelihood for both Bayesians and Frequentists. If the Frequentists were looking for a star and their calculations suggested that the telescope had spotted a fire-spitting dragon in the sky, they would be unlikely to report it. In an email, one of my respondents, a committed Bayesian, put it this way in respect of the search for gravitational waves:

> I think the criteria for the first detection are mostly sociological. It's the level of evidence and group credibility we need to convince people we have seen something entirely outside their experience (a GW source) rather than something entirely within their experience (detector noise, aeroplanes etc). What level that is depends mostly on the attitude of those we want to convince, and their prior predisposition to interpreting the data as gravitational, and less on the data themselves. Once we've done that, GWs are magically within their experience. Life gets easier, and we are free to act like normal astrophysicists (i.e., speculate wildly and mess up a lot without being chastised for it!).

Prior expectations play a role in far more mundane ways too. For example, the claim that there is a coincidence between two signals on two widely separated detectors and that there are only a limited number of coincidences in the background depends on a prior model of a gravitational wave in which the wave comes from a limited area of the sky and travels at the speed of light. Given that this is the velocity, one knows that the two components of a "coincidence" between two detectors separated by two thousand miles cannot occur more than about 1/100th of a second apart. That means that when one is trying to work out the background of coincidences that might be caused by noise alone, one can ignore all pairs of incidents that are separated in time by more than 1/100th of a second—those aren't "coincidences." If gravitational waves traveled at the speed of sound, one would have to take into account all "coincidences" where the two events were separated by three hours or less, and one would be able to rule out background coincidences only where the time of impact of the two components was more than three hours apart. In this case, gravitational wave detection as we know it would be impossible. And yet it has never been experimentally or observationally "proved" that gravitational waves travel at the speed of light—that is part of the point of the very science that is being described on these pages. In still more detail, the same circularity applies to any attempt to detect gravitational waves that depends on a template or even, as in the case of the continuous sources and the stochastic background, on a rough model of the source. The effective prior models of the Frequentists have also been at the heart of the criticism of all of Weber's and the Rome Group's positive claims.

The Bayesians simply say that all these prior expectations should be an explicit part of the calculation before you complete it, and they should be represented by a number. The Frequentists prefer to disown unlikely results "after the event." Mostly, but not always, Frequentists and Bayesians come to the same conclusions at the termination of the calculation, but the "not always" can be important, and is important in this story.

The religious war continues because the Bayesians believe the Frequentists are throwing away or misrepresenting valuable information and/or concealing their use of it, or using it as a post hoc decision-making mechanism. On the other hand, the Frequentists point to the fact that it is very hard to put a number to the "prior probability" of something being right; to put the prior information into number form is try to disguise "subjective" guesses as "objective" information. The Bayesians say that it may be a guess, but at least everyone can see what the guess is, and they can criticize and make their own guesses if they want. The Bayesians say that everyone agrees that miracles (unexpected results) require extra evidence if they are to be proved and that their approach makes this requirement part of the explicit procedure, whereas the Frequentists just agree that something is implausible with a nod and wink and then move on.

The Subjectivity of Frequentist Statistics

Joe Weber's practice of "tuning to a signal," discussed in chapter 3, has already shown how frequentist statistics can be subjective. Every twist of the knob counts as a separate look, or "cut," of the data. If the unlikelihood as reported in the publication is 1 in 10,000 that the result was due to chance, then, if there have been two cuts, it is really 1 in 5,000; if four, 1 in 2,500; if ten, 1 in 1,000; and if there have been one hundred cuts, it is 1 in 100. If there have been many cuts and the paper makes no mention of them, then what looks like a surprising and significant result can be simply a consequence of what is known as "statistical massage" and of no physical interest at all. If all this is done deliberately, then it is cheating. If it is done inadvertently, then it is just bad statistical-craft practice. Immediately one can see that to understand a statistical result, and that includes a frequentist result, one has to know what was happening at the bench and at the computer keyboard from long before the paper was finally written up. One has to be, in other words, a perfect historian if one wants true objectivity. One has to be a perfect judge of character if one thinks one should take this but not that scientist's word for what they did, and, even if one's assessment of their trustworthiness is flawless, one still has to rely on their own understandings and their memories of what they did. This is part of what is meant by "hidden histories" in the title of this chapter.

But that is not the end of the history. What also matters is what happened outside of the laboratory, as can be illustrated by the case of parapsychology experiments. It is a embarrassing fact that there are an awful lot of experiments that seem to show that, for example, a person trying to use psychic powers to guess the images on cards that are being looked at by someone in a remote location who is trying to "transmit" the image tends to guess right slightly more often than they should if the result were pure chance. The statistical significance of these experiments is generally better than those thought publishable by psychology journals, and the design of many of the experiments seems sound even after the most painstaking examination by determined critics.[2] The resolution offered by some of the more honest but determined critics is that even if the unlikelihood of the result being due to chance is 1 in 1,000, for every positive experiment

2. One should ignore those critics and skeptics who find ever more ingenious ways to explain how a protocol *could have been* violated; no science can stand up to that approach.

that has been reported, 999 have been carried out which obtained null results (or negative results), and they have simply not been reported—they have been consigned to the "file drawer." If this is the case, the one positive result is entirely unremarkable. Only positive results are reported because only positive results are interesting, but they are nullified by the "file drawer problem."[3] For the file drawer to be a genuine problem does not imply that anyone is cheating. Rather, the point is that to know the objective meaning of a positive result one has to be a perfect historian, this time not only of the individual scientists' past experimental life but of the past experimental lives of all the other scientists doing similar work. Only when all this is taken into account is there a chance that the right number will be used in the judgment of the degree of unlikelihood that the result was due to chance.

The file drawer problem affects physics too. High energy physics is said to have moved, in the 1970s, from a 3 sigma level of significance to higher levels as a result of such considerations. Allan Franklin writes of this as follows:

> Thus, [in the 1960s and 1970s] the observation of such [a 3 sigma] effect was evidence for the existence of a new particle for any one experiment. But, in fact, the data implicitly refer to a sample space containing a much larger number of experiments. . . . So, quite correctly, [Arthur] Rosenfeld argued that one should not consider only a single experiment and its graphs, but all such experiments done in a year. This made the probability of observing a 3 [sigma] effect considerably larger. Changing the criterion to 4 [sigma] lowered that probability considerably. (Franklin 1990, 113)

According to Franklin (private communication), the 4 sigma criterion was the norm by the late 1970s.

But Franklin (and we must assume Rosenfeld) still don't get to the heart of the matter. Why pick a year as the boundary of the sample space? A year is entirely arbitrary. The sample space is all experiments of that type that have ever been done. And that, of course, leaves open the question of what is an experiment "of that type"? Furthermore, why was 3 sigma

3. Incidentally, the parapsychologists argue that so many positive experiments have been done with such high significance that even if everyone had been doing negative telepathy experiments since the start of civilization and leaving them in the file drawer, it still would not counteract the positive conclusions.

counted as satisfactory in the first place and why was 4 sigma counted as satisfactory afterwards?[4]

Nowadays, and we will return to discuss the matter further, high energy physics takes 5 sigma to be the publishable level. I asked Jay Marx, the current director of the LIGO project, and himself an ex-high energy physicist, why this was.

Collins: How did 5 sigma come to be established in the field that you come from?

Marx: Years ago difficult experiments were done to study the weak interaction. Some of those experiments with published high significance—3 sigma and greater—later turned out to be wrong, while experiments that had a 5 sigma effect mostly turned out to be right. The result was a common wisdom—or mythology—that one should not be confident in a result unless it was a 5 sigma effect. I was taught that as a student. It is because difficult experiments can be subject to unknown systematic errors. When you have a quoted confidence level, it assumes you know all your systematic errors, which may not always be true. A 5 sigma confidence level seemed to give one enough confidence because it gives a wide enough berth to cover the unknowns.

4. For an early discussion of how certain statistical result can mean different things to different people see Pinch 1980. There continue to be statistical disputes and ambiguities in the most technical elements of contemporary science. Here is an email circulated by one of the more accomplished statisticians in the LIGO collaboration (the issue continued to be debated for some time):

> I talked with 'X,' one of the BaBar statistics gurus, who is a confirmed frequentist, but I think in the end we agreed that it doesn't matter whether you are a frequentist or a Bayesian (as far as I'm concerned, if we're writing down and integrating over probability distributions for the distance to M31, we're Bayesian).
>
> But the main point is this: When setting 90% CL *upper limits* on something (like an efficiency, for which we can define a probability distribution), the ONE-SIDED interval [e90%, 100%] (or [0, D90%], resulting in the 1.28 number for Gaussians) is what people in high energy physics assume you to mean. When I described the approach of taking the 2-sided interval and choosing the worse number (resulting in the 1.6 number for Gaussians) he replied (I paraphrase): that's conservative, wrong, crazy. He never heard of doing that for a X% CL upper limit, and didn't understand why anyone would want to do that (and I couldn't enunciate an argument because I don't understand it, even after reading Patrick's note saying that approach "is the most natural.")
>
> Of course, I'm not saying that just because people in HEP do things one way, we should too; but HEP has been setting 90% CL upper limits since before I was born (and before anyone knew the difference between frequentists and Bayesians).

Collins: So the actual number of 5 just grew up as a tradition in the field as a result of experience?

Marx: Right: We're saying we don't believe it unless the significance is extremely high because there may be more uncertainties than are reflected in a published error.

The problem of mistakes is, probably, still more severe in the social sciences, which typically take a 2 sigma level of significance as the publishable standard. Two sigma implies that, other things being equal, a result might be wrong five times in one hundred. There is no rationale that I know of for the different significance levels in different sciences except what can be accomplished in practice. There seems, therefore, every reason to suppose that a large number of social science results are wrong, especially as social scientists seem generally unaware of the problems and are not particularly careful about using many cuts (or tunings) until statistical significance is achieved, nor do they think about or declare the process by which published results are selected from the body of un-publishable analyses.[5]

The Trials Factor

Even when everyone is self-consciously aware of the dangers of the processes just discussed, it does not mean that the problem has gone away. When they are self-consciously worrying about it, physicists refer to the problem as "the trials factor," and it is a real concern for LIGO data analysis.

The organization of the work has a bearing on the trials factor. In chapter 2 the four groups that are responsible for looking for different kinds of signal were described. The Equinox Event, it will be recalled, was spotted by the Burst Group. A member of one of these groups, whose anonymity had better be preserved, good humouredly described the different characters of the competing groups in terms of categories from the TV series, *Star Trek*. This was in July 2007, before the Equinox Event had appeared. The Inspiral Group, he said, are like the Borg. According to Wikipedia, "The Borg is a species without individuality where every member is a part of 'the collective' in an attempt to achieve perfection. They assimilate species and their technology when it suits them." That is, they are very efficient and hard work-

5. A biological scientist with whom I entered into casual discussion told me that in her field 2 sigma is also the norm and that they expect about 50 percent of published results to be wrong.

ing, strongly and authoritatively led, and continually expand their activities. This style of organization, I would suggest, is appropriate for a group that must organize a search through a massive series of wave-form templates.

The Burst Group is in contrast chaotic—my respondent likened them to the Ferengi, which Wikipedia describes as follows: "They and their culture are characterized by a mercantile obsession with profit and trade and their constant efforts to swindle people into bad deals." The point is that they are much less well organized, without strong leadership; each member of the group insists on doing things his or her own way: There is a "bazaar" of competing different methods all running in parallel. The inefficiency but creative freedom of the bootlegger is typical of their approach, perhaps made necessary by the quintessentially unknown properties of the bursts they have to look for.

The Burst Group, then, has split itself into competing factions, each doing their analysis in their own way. Incidentally, the term "pipeline" has come into use for a method of data analysis that culminates in a result associated with a statement of statistical confidence. The Burst Group, then, has within it a number of groups each using a different pipeline, all of which they have developed in competition with one another. This is excellent from the point of view of data analysis creativity and cross-checking, but it creates a trials factor problem. When analysis groups proliferate, it looks as though the significance of the results of any one group may have to be divided by the number of different pipelines all looking for the same thing in different ways.

Here is how one member of the community put it, referring not just to what was going on in the Burst Group but to what was going on across the whole data analysis collaboration:

> I just want to say that, if you look across the collaboration as a whole, we have four different search groups, each of which is running a handful of different analyses, so there's something like ten different analyses running—more than ten analyses running across the collaboration—so if you want to say something like we want to have below a 1 percent mistake rate in our collaboration, that means, right away, that you need false alarm rates of something like one in every ten-thousand years if you want to be sure. It's a factor of ten because there's at least ten different analyses running across the collaboration.

But this is swampy ground. If it is the case that the significance of a positive Burst Group result is affected by a negative stochastic background result—the result of a group looking for an entirely different phenomenon except that gravitational waves are involved—why stop there? Why isn't a positive Burst Group result affected by experiments being done in a completely different branch of physics? Or what if someone, somewhere, is stealing LIGO burst data and doing other kinds of analyses on it, all producing negative results, yet unknown to the collaborators? Does this mean the original results are vitiated? This kind of problem was already hinted at when the Rosenfeld argument for raising the standards of high-energy physics from 3 to 4 sigma was discussed—the boundaries of the appropriate sample space are vague.

Can I sabotage a positive result by doing some quick negative runs on the same data? One scientist put the point very neatly:

> That's the other problem, there's a bunch of [different] numbers associated with this. . . . Somebody . . . wrote a pipeline . . . that has never been looked at or studied, which says once in three hundred years. I could write a pipeline tomorrow, I'll bet, which could not see the Equinox Event, thereby degrading its significance. And all I have to do is write a really bad pipeline. In fact I can write eight bad pipelines that are bad in different ways so they're all uncorrelated, and when I do that all eight will miss that event, that will downgrade its significance immediately. So if I wanted to I could kill that event because of the trials factor.

As this respondent points out, the trials factor problem occurs only if the methods are "uncorrelated" or "independent." Must one, then, actively discourage other people from analyzing the data in independent ways in case they reduce the statistical significance? And, as the scientist who made the remark immediately above also pointed out to me, reinforcing some of the other examples, whether a second result detracts from a first result or adds to it may depend on exactly how it is done. In this case, one of the pipelines actually did produce a positive result while a second actually did produce a negative result; the implication drawn by some was that the significance of the positive result should be divided by two. But, as he pointed out, if, in the case of the "negative" result, the threshold for what counts as "positive" had been lowered only very slightly, it too would have turned out positive, not halving the significance at all. Something weird, or at least indeterminate, is going on here. As my respondent put it: "[There can be situations] where the whole trials factor argument is meaningless—it's just stupid."

The discussion of the Bison group's 1-in-300 year claim illustrates the trials-factor problem in day-to-day practice. It was said the Bison group had tried twelve different ways of matching coincidences, of which only one had delivered anything remarkable. The 1 in 300 should be divided by 12, giving 1 in 25, which was in the same ballpark as the existing group consensus. Bison's defense was along the lines that the twelve trials were not independent, so there was no dilution. This defense, as it happens, made no headway with the other physicists but indicates the extent to which these questions are debatable.

To summarize: to take the trials factor into account, one must first decide what counts as *the same experiment*, and only "the same" experiments count in the calculation. Experiments that are different do not count. Then one has to decide on the boundaries of the social and temporal "space" which will be searched for experiments of the same kind. Then having collected the set of experiments of the same kind, one has to decide which trials within each experiment were independent and which were correlated, and here is it only the "independent" ones that count. Even if these qualities—"sameness," "in the right space," and "independence"—could be defined, they would still come in gradations, not in "yes's" and "no's."

And even if one did know, logically, as it were, the exact right way to define all these imponderables, how would one gather the facts of just how many trials had been conducted in one's immediate location and elsewhere? Once more, as well as solving all these quasi-philosophical problems, it seems one needs perfect knowledge of activities of all the actors in the world who are potentially involved if one is to resolve the number that correctly represents the confidence one should have in a conclusion.

What Did You Have in Mind?

If only it were so simple! But there is yet another layer of difficulty: the meaning of a statistical result depends on what was in the minds of the research team. Suppose I ask for your birthday and you say "July 25th." I say "Amazing! That was the date I had in mind, and the odds against that are 365 to 1." Well, if I did already have that date in mind, the odds are indeed 365 to 1. On the other hand, if I did not really have it mind, there is nothing of any interest going on.

To see how this works out in gravitational wave physics, we can go back to "the Italians" and their 2002 paper. It will be recalled that a central claim of the bitterly received 2002 paper was that the two bars registered an excess of coincidences at a regular part of the sidereal day. In other words, over the course of twenty-four hours there was one hour that registered a peak of activity and, perhaps, another, twelve hours later, that registered a very small peak. The second peak would be expected given that the Earth is transparent to gravitational waves, so that the orientation of any one bar detector in respect of the Galaxy effectively repeats every twelve hours, but it could have been worrying that it was not more marked.[6]

As mentioned, a crucial criticism of this finding was made by an American analyst, Sam Finn. He said that he had worked out that if you take a purely random distribution of events, in which the number of events is equal to the total number of events reported by the Rome Group, and divide them up into twenty-four bins in two different ways, there is a one-in-four chance that the two resulting distributions will differ by the amount in the Rome data. The question being asked here was whether the data justified the claim that the peak was correlated with the Earth's relationship to the Galaxy—the sidereal day—rather than with the Sun—the solar day. Finn's point was that the fact that there was a peak when the data were analyzed according to the sidereal day and not when the data were analyzed according to the solar day was statistically unsurprising and represented virtually no gain in information: if you modeled the procedure with random numbers, you could get a difference in event rate of this size one time in four.

Furthermore, Finn argued, even the original claim that there was a zero-delay excess in the first place, irrespective of the clustering, was statistically insecure. He calculated that, in any random distribution of Rome-like data over twenty-four bins, the statistical unlikelihood of one bin standing out to the extent that had been presented was equivalent to only a little more than one standard deviation—again about one chance in four.

In their paper, the Rome Group had reported that the likelihood that the clustering that they had found would be due to chance was only 1.35 percent. How could there be such a difference between their account and Finn's?

The explanation of the difference is simple but revealing. The Rome Group's level of significance was justified if the analysts had set out to look for a peak associated with a *particular hour* during the day rather than

6. The business of twelve versus twenty-four periodicity came up in the early days of Weber's work (see *Gravity's Shadow* and chapter 1, above). It now seems that the analysis of "the Italians" shows that a twenty-four hour periodicity is acceptable, even though Joe Weber's finding a twenty-four hour periodicity was taken by many, and subsequently by Weber himself, as being an impossible result. In retrospect, Weber's initial finding might have been acceptable, but the fact that it seemed to change in the face of criticism was not!

a peak in any unspecified hour. Finn's calculation that a likelihood of 27.8 percent was for *any hour* turning out to have such a peak of coincident events.

Now one can see the problem for understanding the meaning of such a statistical claim. If the Rome Group had done their analysis and only later noticed that the hour containing the peak happened to be one where the detectors were most sensitive to the Galaxy as a source of signals, then their procedure would fit Finn's metaphor—set down in his published reply—of firing the arrow first and drawing the bull's-eye later. This would be a clear case of post hoc statistical massage, the equivalent of my saying I had your birthday in mind before you mentioned it even though I hadn't. But if "the Italians" had set out to look only for peaks in the hour when the detector "faced" the Galaxy square on, then their 1.35 percent chance calculation was correct—the bulls-eye would have been drawn prior to the flight of the arrow—as when someone really has the guessed the date of someone's birthday before they tell it.

Someone, let us call him "X," told me that he believed "the "Italians" had not chosen that hour in advance. X reported to me a conversation he had had with a member of the Rome Group in which it had been confessed that they would have reported the result irrespective of the hour in which the peak had occurred. And there is a rationale for this position. As it was, there was some problem about whether there was one peak or two in the course of twenty-four hours. If the Galactic center was the source, then two peaks would have been expected. But the one peak could be explained by the particular orientation of the two detectors if the source was not the center but the whole Galactic disk and the disk was put forward by the Rome Group as the source. Yet the Galactic disk is not the first source one might think of because of the concentration of stars at the Galaxy's center. Furthermore, the Rome Group had, at different times, speculated about other sources, such as a "halo" of dark matter surrounding the Galaxy, which would justify alternative directions as potential sources.

On the other hand, the Rome Group defended itself against accusations of any such post hoc data selection in a paper circulated on the electronic preprint server:

> The Galaxy is certainly the privileged place of the sources attainable by present GW detectors and we think that the experiment described . . . should be considered as based on the "a priori" hypothesis of signals originating in the Galaxy. This was clearly indicated in [a] previous published paper: . . . "No extragalactic GW signals should be detected with the present detectors.

> Therefore we shall focus our attention on possible sources located in the Galaxy." (Astone et al. 2003)

On other occasions, however, such as in their published defense of the 2002 claims (see below), they seem less concerned with evidence about the timing of statements about prior expectations, suggesting that it is legitimate to consider any models that seem sensible, whether developed before or after the analysis, so long as they are not *chosen* to increase the saliency of the results. This claim is justified by a Bayesian approach, which will be discussed shortly. But as far as the frequentist analysis is concerned, one can see that it points straight at the motive behind the choice—which is the *internal state* of the analyst when the choice was made.

The sociological point is this: We would be ill advised to try to work out what the Rome Group had in mind—the discovery of persons' internal states is a perilous enterprise, and for the sociological purpose at hand there is no need to try to discover them or even speculate about them.[7] Nevertheless, it remains that for the purpose of frequentist statistical analysis it is vital to know the internal states, and this is made evident by the amount of time and energy that the physicists have expended in trying to establish what they were. X, who was an accomplished statistician, treated the spoken report of what the Rome Group would have done had the clustering hour been different as highly germane to what was being claimed. And, Bayesian analysis aside, in concrete terms, it amounts to the difference between a 1 in 4 likelihood of the 2002 results being due to chance—a result almost certainly not worth pursuing—and a likelihood of 1 in 75—something that one might well want to follow up.

The way this dilemma is normally resolved is that analysts are expected to state what they are looking for in advance, as Astone et al. claim that they did. For now, let us sum up what has been established. In spite of the vaunted objectivity of frequentist statistics, to know the true meaning of a statement of probability in a published paper, one must know the history of the reporting teams activities, and one must know the history of everyone else's "similar" experimental and analytic activities. One must define a time period, and a set of physical locations, over which one is going to count activities as "similar," and one must work out what "similar" means and what an "independent trial" of a similar experiment means. Then one must know what the analyst had in mind, which is the sort of thing that is

7. Again, this is the Anti-Forensic Principle.

normally defined only with a considerable degree of uncertainty, and often in the face of determined disagreement, after a prolonged legal trial.

The Bayesian Approach

What I have tried to establish in the last section is that frequentist statistics are full of subjectivities and irresolvable uncertainties. This counters one of the arguments used against the Bayesians—that frequentist statistics are objective while Bayesian statistics are subjective. Actually, both are subjective, and both make it hard for the uninitiated to recognize that subjectivity by presenting the conclusions to their analyses in the form of outcomes of calculations. The Bayesians proclaim that putting a number on one's prior beliefs forces one to state them clearly and that this is a virtue.

The case for the Bayesian approach is most easily made, once more, by thinking about the approach of "the Italians." In a long and technically dense paper published in *Classical and Quantum Gravity* (*CQG*) in 2003, Astone, D'Agostini, and Antonio offer a Bayesian defense of the 2002 paper by Astone et al. They argue that the meaning of the 2002 paper depends on one's prior expectation—that is, on the number of gravitational wave events per year that are thought reasonable. If that number is very high, then the paper shows that the expectation is wrong. If the number is very low, then the paper provides no new information. But if the prior expectation is in the region of what was apparently found, then the findings give strongest support to the Galactic center as their source (interestingly, under this analysis, not the Galactic disk). They do not claim to have proved anything beyond this—that is, they say, not they have definitely found gravitational waves, but only that their findings add something in the way of information in the case that their prior assumption about the expected rate, marginally backed up by the actual findings, is correct. They claim that this is how science works—by adding one small piece of information to another. And it seems reasonable to say that since even the frequentists have to use prior expectations, or models, if they are to do any science at all, changing the pattern of priors even in such a small and provisional way cannot be a bad thing.[8]

8. Astone, D'Agostini, and Antonio's long Bayesian paper, I should add, had no impact whatsoever. Google Scholar shows only one self-citation. Here, then, I am engaged in the rather peculiar exercise for a sociologist of treating seriously a paper which the scientists themselves treated as "invisible." I believe my license to do this lies in the fact that I am not using it to support the 2002 paper—which is what the scientists were interested in—but to add something to a much more general point about the gradualist nature of scientific discovery.

It has to be said that the position of "the Italians" has not been completely consistent through these twists and turns. At one time they claimed that it had been clearly stated a priori that the Galaxy was the preferred source, but even then there have been switches between Galactic center and Galactic disk. It could also be argued that a confirmed Bayesian would have kept quiet about the whole thing, given the very low prior probability of seeing anything much with detectors that were so insensitive according to prevailing theory and also given the context of consensual astrophysical opinion about the nature and prevalence of potential sources. It is very non-Bayesian to talk of "the experimentalist's right to look at the world without theoretical prejudice." (It should be borne in mind that I said that—it is my phrase. I did not hear any of "the Italians" actually say it, though the sentiment is clear enough in the quotation from David Blair [see chapter 3] and in much of the discussion of these events in *Gravity's Shadow*.) So it might be said that "the Italians" were Bayesian when it suited them and not Bayesian when it did not. But, once again, I am going to invoke the Anti-Forensic Principle: I neither know nor care about any of this because it is not my job. In any case, though it should never be erected as a guiding principle, in practice it may be good for scientific progress to take one position in one context and another position in another context; human beings do it all the time. Here, however, I am interested only in the logic of any consistent position one could extract from the various arguments.

We can now ask whether it is good or bad for science to publish papers such as Astone, D'Agostini, and Antonio's 2003 paper in *CQG*, which make only small claims, and which are only valid if certain initial, optimistic, assumptions are correct. It is obvious that you are not going to want them published if you are championing a much newer and more expensive technology that is based on more pessimistic prior assumptions about the flux of gravitational waves such as to render the old technology obsolete. But let us set this aside and try to consider "the view from nowhere."

Here is a problem for the frequentists: if you demand that a paper should never put forward what one of my American physicist colleagues dismissively referred to as "indicazioni," but only firm discovery claims, then it is hard to establish prior hypotheses. We have seen that, even in frequentist terms, what "the Italians" had in mind crucially affects the meaning of the results—either their clustering was subject to a likelihood of being random of 1 chance in 4 or 1 chance in 75, depending on whether the hour of interest was chosen before or after the results emerged. But the only firm and clear way to establish a prior hypothesis is to broadcast it through publication.

The alternative is a Catch 22. If you have a provisional claim, you should not publish it, but the only way to make your next claim less provisional is by publishing a provisional claim. Only by publishing their claim could "the Italians" put themselves in a position to avoid, after their next round of data gathering, the charge of painting the "bull's-eye" after the fact. And, as it happens, their next round of data gathering does not seem to have supported the now clearly stated 2002 result, so even the LIGO-based frequentists should be delighted. The 2002 publication made it impossible for "the Italians" to shift their ground, should they have wanted to shift it, in order to support a new interpretation of the aggregate data as new results came in. As a result of the fact that new data did not support the clearly stated 2002 claims, we now know that the 2002 paper contained no new information at all—it has disappeared in as clean a way as it is possible for something to go away in science. And what harm has been done? On the other hand, if additional data had reinforced the 2002 claim, it would have had a cumulative importance far beyond that of a brand new announcement of the aggregated data set: a "bold conjecture," as Popper would say, would have been made, wide open to falsification, that would have enhanced the "scientific-ness" of data that confirmed rather than falsified it.

The four grounds for not publishing are, first, embarrassment about the history of the field and its many unsubstantiated claims, which scientists feel make the subject a laughingstock among the scientific community—this is not a "scientific" reason. Second, the interest of the interferometer groups in having the only viable technology—again this is not a ground that can be legitimately announced in public. Third, the view that science like this should be involved only in the binary process of discovery/non-discovery, or Nobel/no-Nobel, in which case "indications" claims are sneaky, back-door attempts to take credit from others who better deserve it—which has to do with reputations and rewards, not knowledge. Fourthly, a preference to present science as a producer of certainty, in which case disputes and disagreements should be kept in-house—something which I will argue in the Envoi is not the best way to do science.

The view that disputes should be kept in-house I have described elsewhere as a preference for "evidential individualism," in which each individual or laboratory is privately responsible for the entire chain of scientific discovery from provisional indication to published observation claim.[9]

9. *Gravity's Shadow*, chapter 22.

Its counterpart is "evidential collectivism," where the wider community takes responsibility for the scientific truth through public debate of openly published provisional claims. That evidential individualism played a part in the rejection of the Rome paper by the community is easily seen in this selection of quotations from the 2002 GWIC meeting at which the matter was debated:

> In particle physics it's traditional that the particle physics community . . . For 99 percent of physics results it goes through a process of vetting inside the particle physics community before it goes to the press and before it goes to an archival journal.
>
> We [should] . . . agree on a kind of set of guidelines of how we proceed in this community and how we present things . . . because we are going to start having results and as we have results we really need a way that we all try to follow—not exactly—the same rule, that we try to follow as we go forward to present these results so that we present the best kinds of paper which contain the best information and not debate them after they're in press—I think that's the worst problem is if you debate them after they're in press. Inside our community it makes even things that can be very right—they can be right or wrong—but it makes the whole thing controversial—unnecessarily.
>
> The question . . . is where should the controversy be? Should the controversy be after a paper is published in a journal and should it take place in the press, or should the controversy take place between the release of pre-prints for more general comments?
>
> The question is when you go public with your data—is that when you publish it or is there a step in-between where you expose it to the community, which is us.

What ought to be published? Should indications of possibilities be put into the public arena or should everything be kept in house until certainty is reached? We'll come back to the question.

This chapter, as well as opening up the question of whether uncertain results should be published, has revealed that even though statistics makes calculations, it is really representing judgments. Statistical tests have histories which affect their meaning but which are never completely knowable. A statistical test is like a used car. It may sparkle, but how reliable it is depends on the number of owners and how they drove it—and you can never know for sure.

6 The Equinox Event: The Denouement

The book began with my sitting in Los Angeles airport waiting for my flight home after the meeting in Arcadia at which "the envelope" was opened. We are now back in Arcadia, at the beginning of that meeting, and I will shift to the present tense.

Though not everyone is agreed that all the proper work is finished, the consensus is that it is time to reveal what was what. The time, from first noticing a promising event candidate to finishing the work on its analysis, has been eighteen months. This seems too long if gravitational wave detection is ever to be a science that can rank with the other kinds of astronomy. One needs a result that can be compared pretty quickly with other astronomers' sightings if the promise is to be fulfilled. As it is, the opening of the envelope has been promised over and over again only for the moment to be put off, as people realized that still more prior analysis needed to be done. Remember, the problem is that no-one wants to do retrospective analysis, and that as soon as the contents of the envelope is known, any further analysis is going to look "wise after the event." So, within reason, people have wanted all the analysis that could be done, to be done, before the envelope is opened. As it is, there have been hold-ups in completing some pipelines and in finishing the full analysis needed before "boxes" could be opened. Though there are people who think necessary prior work is still incomplete, the sheer time that has elapsed is making their view seem unreasonable. It might have been otherwise if the Inspiral Group had hurried their analysis along as fast as they intend to in the future.

The Abstract

The Burst Group have now written their draft abstract. This is the abstract that will be submitted for publication if it turns out that the Equinox Event is not a blind injection. It has to be written because it reports the upper limit results for S5, and, if the Equinox Event is not an injection, then it must figure in the upper limit calculation. The rule is that the papers have to written with conclusions reached, and firmly and unambiguously stated, before the envelope is opened; there will be no backsliding. The Inspiral Group has found nothing of any significance. The Burst Group has to deal with the Equinox Event. There are three possibilities for how it might turn out—injection, correlated noise, or an event which is below the threshold for reporting as a discovery. The concept of "injection" does not belong within the conceptual universe of the paper, so one of the other two choices has to be made. I understand there have been huge debates over what should be in the abstract, the contents being prefigured by the argument at Amsterdam about Antelope's concluding remark. The full abstract of the proto-paper as finally drafted will be found in Appendix 2. The crucial sentences referring to the Equinox Event read as follows:

> One event in one of the analyses survives all selection cuts, with a strength that is marginally significant compared to the distribution of background events, and is subjected to additional investigations. Given the significance and its resemblance in frequency and waveform to background events, we do not identify this event as a gravitational-wave signal.

Those sentences might, instead, have looked something like the following:

> One event in one of the analyses survives all selection cuts, with a strength that is marginally significant compared to the distribution of background events, and is subjected to additional investigations. This event cannot be ruled out as being a gravitational-wave signal, but it resembles background events in frequency and waveform and its significance is too low to justify a positive identification.

This seems a small difference, but it is not; that is why the exact phrasing was argued over for so long. Again, it is worth noting in passing how the work of science is done: there is calculation, but, in the last resort, there are words. The Burst Group has expressed itself in such a way that the most ready interpretation of the event is that it is noise. They have chosen not to say that it could be an event that is too weak to count as a discovery. At least some members of the collaboration believe one of the crucial components is a glitch. Should they have wanted to express themselves in another way, they have been pressed not to do so by strong negative opinions such as those expressed by members of Virgo.[1]

What if it is a blind injection and not noise? If a blind injection, then it was intended to look like a gravitational-wave signal, and the fact that it also looks like noise just goes to show that gravitational waves sometimes look like noise. In that case, has the Burst Group made the wrong choice? Given the weakness of the event, making the wrong choice would not be a serious mistake in terms of physics but it might be a mistake in terms of mindset. The members of the Burst Group will not have allowed themselves to be excited about the possibility that they have seen a weak gravitational wave; they will not be filled with regret that their proto-event did not come up to scratch. Caution will have trumped hope.

Opening the Envelope

The room is full and pregnant with anticipation. Jay Marx, the director of LIGO, is to open the envelope. He stands at the podium, PowerPoint at the ready. He makes a nicely judged and well-received wisecrack that relieves the tension a little: "I've given lots of talks to this group, and this is the first time I've not seen two hundred people checking their email."[2] Marx then shows a jokey slide of a very old envelope, but people are becoming impatient and this does not work so well.

He then puts up his first proper slide. There is a moment's silence as the group takes in its meaning (figure 12).

1. It was members of Virgo who also insisted that the actual statistical confidence associated with the Equinox Event should not be mentioned in the abstract.

2. It is the habit of physicists (one which I have acquired), to work on their networked laptops throughout all such events, whoever is speaking. It does not necessarily seem rude because they could be making notes, looking, online, at the PowerPoint slides being presented, or checking calculations made by the speaker. But mostly they are writing programs, working on their own talks, or answering emails.

Figure 11. Jay Marx prepares to open the envelope

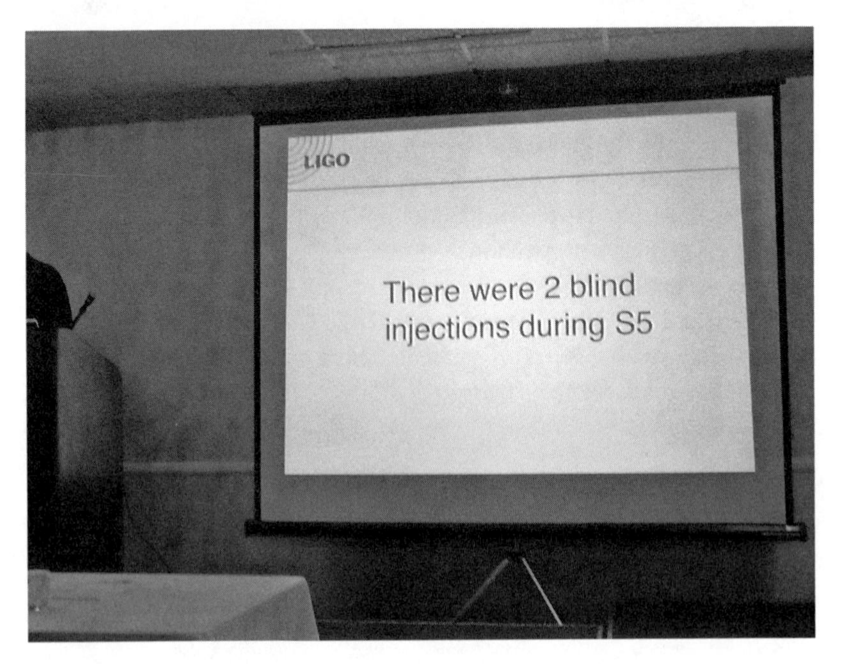

Figure 12. The first slide from the envelope

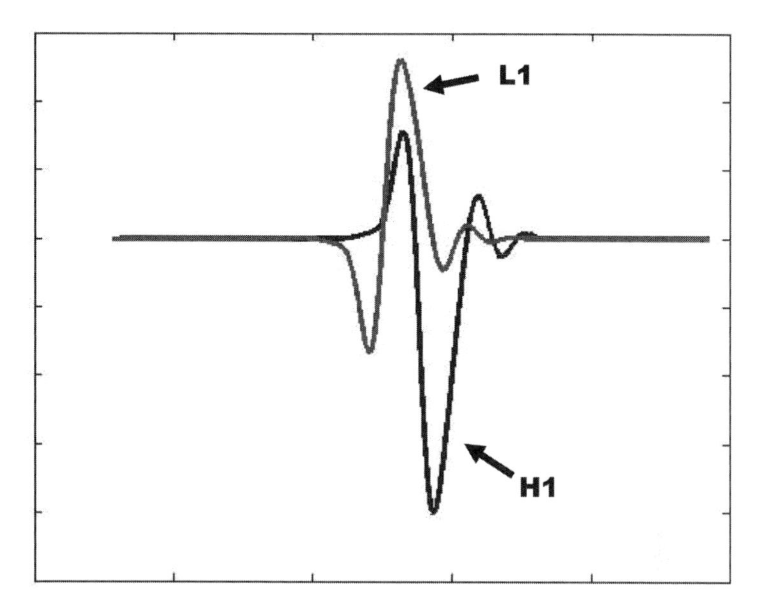

Figure 13. The injection

There were two blind injections in S5! All the discussion had been about the Equinox Event, but there were two blind injections, not one. At best, one of them had been completely missed.

And it had: there was an injection on 13 September of a loud and clear inspiral that did not look like a glitch in any way and that had not been seen at all. The second injection was the Equinox Event, and the Burst Group had nearly exactly identified its characteristics before they decided it was too much like noise. The injection team's input into H1 and L1 is shown in figure 13.[3]

Marx says: "We observed something significant [inaudible] for one of them. But we didn't have the chutzpah to say anything."

There followed a strange few minutes as the audience came to terms with what they had heard. A lot of people, including me, had expected this envelope opening to be an anticlimactic event. It was felt that it would anticlimactic if the Equinox Event was not a blind injection because it would just be the noise that everyone thought it was, and it was felt that it would be pretty uninteresting if it was a blind injection because "so what?" But

3. Comparison with figure 4 shows the match between injected and extracted signal and illustrates the ever-present noise in the detector and the influence of its response-function on what is seen.

Figure 14. Aftermath of the Envelope Opening—about one third of the room

actually, the fact the Equinox Event was a blind injection did not turn out to be uninteresting at all when the moment came—that it had not been treated with more respect suddenly felt like a matter of real concern. This is not to mention that there was a second injection that had not been seen. The audience reaction was a mixture of the jokes and laughter that relieve tension, and that Marx had licensed in his introduction, interspersed with moments of quiet stillness purporting something serious. My note, made at the time, and referring to the Equinox Event, reads: "The room is really quite silent. People are wondering why they didn't go for it."

The next half hour was fascinating as people came to terms with what they had heard and began to stake out positions. We will concentrate on the Equinox Event, as the other injection—the missed inspiral—was an altogether simpler matter, though one about which regret was expressed during the discussion by a senior member of the Burst Group:

> This is vetoed by a category 3 veto because there was high micro-seismic . . . but it is a category 3 veto and because we want to analyze data—because we want to detect things in that data—and we have not produced that histogram. These injections might have been gravity wave signals and before we

> opened the box we forgot, I forgot to ask for those histograms, which I often do. . . . We always said we will look in all the data we analyze and we didn't look at all the data we analyze.

Within a few moments a member of the Inspiral Group had found the inspiral event by analyzing the appropriate stretch of data on a laptop and had shown how the analysis algorithm correctly highlighted the right template with the right masses of stars, distances, and so forth. It had been missed because, as the quotation indicates, it had occurred in a stretch of data vetoed by data quality flags. The vetoes, however, were category 3—the light touch vetoes that invite a further reexamination. That reexamination would have happened in due course, but they had not done it prior to the opening of the envelope. Both an in-between procedure called "looking at the histogram" and another which involves looking for coincidences between H1 and H2 would have indicated that the vetoed data needed more attention. Both were meant to have been carried out alongside the main analysis, but they were not done under the pressure of completing the main analysis in time for opening the envelope. The event would have been detected eventually because those stretches of lightly vetoed data would have been reexamined. The incident illustrates the logistic choices to be made where vetoes are involved. The lesson learned would bring about a reordering of priorities. There is nothing more to say about it.

The Equinox Event is more complicated because the outcome could still be made out to be a success. If the event was not clear enough to count as a detection, then the LIGO community could be said to have demonstrated virtue by refusing to give the coincidence credence beyond that which they would give to any other chance concatenation of noise. The first reactions were indicative of the argument to come. The senior theorist who often invokes the trials factor problem spoke up early and forcefully to make his usual point. He said that it was impossible to claim a detection for an unmodeled event unless there was a correlation with something else in the electromagnetic spectrum or the like. This received a robust response from the floor: "We don't need a model. If we see something that is statistically significant, we should have the guts to say so."

Surprisingly, a very senior scientist known for his caution, especially when it comes to confusing signals with artifacts in the devices, said:

> Maybe [I would] argue with [the senior theorist] about this. The only reason for us hesitating was the discovery that there were events [glitches] that looked very much like this in the data. That's the only reason. But the fact

> is that none of them—and if you look at the rates for them at the most optimistic level, you would have gotten rates that were still smaller than what we've now got. . . . People asked the question, would you publish this event, and then I would have argued. That was a completely different discussion. The discussion would then have been: should we wait until the middle of S6 and we see more events? . . . On the other hand I think that the thing that made us hesitate is that many of the events that we look at as background looked similar to this. I think that was very emotional and, OK, I feel guilty.

Still more surprisingly an analyst who had done much to destroy the credibility of the Rome Group's 2002 paper remarked:

> I want to make a comment at a somewhat higher level here. We are, right now, and in the foreseeable future, working at the edge of detectability. That is, we may get very lucky and have a very loud source that we can believe in unambiguously, and, you know, some people will go out and buy their tickets to Stockholm and the like even before the paper is published. But for the most part we are almost certainly going to be in a case where we are not going to have the kind of confidence that some of us would perhaps like to have, and I think we need to get used to the idea that we may have to, as a group, say we have seen something and put ourselves on the line over it. And that is not necessarily a bad thing. But I think we do have to get ourselves into the mind frame that we could be wrong. One of the advantages, however, in astrophysics is no one will ever know. [Laughter.][4]

This remark was echoed by another audience member:

> I was going to say something that sort of parallel to what [. . .] said. Right now we have a conservative mindset, that is, that we don't want to publish that we saw something when we didn't. And I think there's nothing wrong with that, but I think we have to decide are we willing to live with the possibility of not seeing something that is really there in order to be conservative or not. So it's a sort of cost-benefit analysis. . . .

In these last three comments I thought I heard the coo of "Italian" pigeons coming home to roost.

4. It would be right to treat the remark about astrophysics as the joke the audience took it to be.

A senior member of the collaboration, expressing a Bayesian sentiment, said:

> [O]ur conservatism will diminish. When we are looking at our fifty-seventh detection, we will be less conservative. However, we are always going to be at edge of sensitivity for lots of events. So the question of digging signal out of noise is always going to be with us—not for every event but for lots of our events.

Another audience member regretted how things had turned out:

> I am very afraid of what this means. . . . I think . . . the Burst Group burst searches should have been considered more seriously. Maybe not be ready to go out and tell [the world, but] we should certainly have considered it as a detection and we didn't. I think the psychology of many people has [been] "it looks like a blind injection so let's not worry too much and we'll open the envelope." I was hoping it was not a blind injection. I really thought it was real. I didn't think we could claim it [as a discovery], but I didn't think it looked like the background at all. [And some exchanges later:] I don't think it was taken seriously by all of us. *We didn't talk about it in the corridors. . . .* [my emphasis]

Another senior member of the collaboration said: "[The last] point is a very, very serious one. We have not yet gotten ourselves into the mind state where we can detect something."

The final comment, or near-final comment, from the floor is one we will return to. Someone said: "Are we too detection oriented? In other words, why do we need to make a detection in S5 or in S6? What is the rush? . . . What is the scientific basis for rush?"

The Immediate Aftermath

The next couple of days were intense. Some people became very upset. A highly respected member of the collaboration (HRM) was said to have said during a meeting that the blind injection challenge showed that the collaboration was a failure and would never detect gravitational waves. HRM is sure, however, that all he intended to say was that the "exercise showed that the collaboration was afraid to make a claim of detection based solely on GW data AT THAT POINT. . . . [The collaboration will make a detection but] was farther from that point than I would have hoped at that time" (private correspondence, September 2009).

The febrile quality of the discussions is shown by the fact that those who had been responsible for the downgrading of the Equinox Event felt they were being attacked and were very unhappy about it. I interviewed the highly respected member (HRM) during the meeting. He said to me:

> I'm actually quite disappointed. I think more than anything else it shows that we haven't as a community really gotten beyond, I think, an unreasonable fear of being wrong about a first detection. Unfortunately I think everyone's perceptions are colored by Joe Weber's experience, and everyone wants to be so super-cautious. If you just look at the raw statistics, and I'll . . . accept the Burst Group's [first] assessment that it was a 12 percent chance that it was chance coincidence [here the very senior scientist is taking a very conservative view of the statistical outcome], that means it's an 88 percent chance of its having been caused by something. So roughly 8:1 in terms of the ratio of what the likelihoods were. And that's obviously not high enough confidence to make a claim of something new—I'll accept that—but in your heart of hearts you really ought to be believing, you ought to go with the odds, and we didn't do that. . . . The Burst Group actually carefully looked at it and really tried to understand what the systematics were. They did it, however, without making the best use of an experimental understanding of how the detector operates. I don't think they really ever got around to spending much time with the instrumental idiosyncrasies, where they might have found ways of either increasing or decreasing the probability of it being real.

The last remark implies that something might have been done to work out ways of lowering the background through a closer examination of the physical causes of the glitches, which may have enabled them to eliminate some of them. The interview was to take some interesting turns:

> Collins: So you didn't like [the Burst Group paper] because it was too dismissive of the possibility that this was a gravitational wave?
>
> HRM: Yes it was.

The HRM then showed me a long paper produced by the Burst Group analyzing the Equinox Event and revealing that even though it looked like a glitch, its waveform would be what would be expected from certain real events after they had been molded by the "detector response function":

C: Yeah—I have heard people say that the instrument acts as a band pass filter, so what starts looking like an inspiral can end up looking like a glitch.

HRM: Yeah—because all of this stuff [the early part of the inspiral waveform] is at too low a frequency to show up, and so all that comes through is this last little bit, so obviously it doesn't match perfectly with the thing. But why should it? [given that we don't yet know for sure what such a waveform should look like as we haven't proved it experimentally and we have not calculated all the details].

And you know there are things like the reconstructed position that turns out to include the Perseus cluster—a super cluster. Which is one of the largest relatively nearby clusters of galaxies at a distance where this kind of an event, they actually reduce this signal by a factor of six to put it into the noise here.

So there are good reasons for you to say "it might not be real"—they put all those into the paper. There are equally good reasons why you might say "you know, there are some things that make it look pretty good." And those ought to have been in the paper also, and they weren't.

C: I think people have painted themselves into a corner, partly because of the heavy criticism of Joe Weber, and partly because of the really, really heavy criticism of the Italians.

HRM: Yes. There's that also.

C: They were never allowed to say anything tentative—you've either got to claim gravity waves or not claim gravity waves, otherwise you are illicitly trying to get a share in the Nobel Prize.

HRM: And [names another very influential figure] is one of the strongest from that position; he was very harsh on the Rome Group. I read what they did. I think statistically they fudged things a little bit. They didn't include the trials factor in it. But I didn't think what they said was so bad. They said we cannot rule out that this is a gravitational wave. It has some marginal significance, and we cannot rule out that it is a gravitational wave. Well, you probably could have actually from the point of view of the energy density and waves, I guess that's a . . .

C: But that's an astrophysical argument.

HRM: That's an astrophysical argument, and I think it's actually one that's very powerful, by the way, but beyond that I thought that what they had was not unreasonable. And of course the first evidence that we're going to have is going to be marginal evidence. It's going to be at the edge. . . . [These comments very lightly edited by HRM.]

Another very senior scientist who was known for his cautious approach told me:

[T]o me the whole thing smells of "my god, are we going to demand before we start investigating seriously the candidates a perfection in the signals?" I'm not talking about publishing—please don't get that wrong. We have not taken things seriously, and things don't come forward through the collaboration. And so effectively what we've done is made our sensitivity a factor of two or three less good by the act of being so conservative. And that's scary. And that's my problem.

C: Do you think it will change now?

Well, people are going to be mad as hell about this and so things are going to change. I hope.

The clearest version of the other view I gained from discussing the actual wording of the abstract of the straw-man Burst Group paper with a scientist who had been heavily involved in its drafting:

The point that I wanted to make was that a number of people, including myself, did not want the sentence to read like that Rome paper from whatever year it was—2001 or whatever—where what was infuriating about that was its coy playing with "we don't really have good evidence but we want you to think maybe." And that paper haunts us. And people draw different lessons from it, but what haunts me and some other people is not wanting to be accused of trying to straddle the line of winking at people when we aren't actually able to stake our honor on it. So we do not identify this—trying to stay on the sober side of coyness and winking. And I should say, that even having this much about the Equinox Event in the abstract was a heated debate, where a group of our colleagues, mainly the Virgo ones, felt that this was giving way too much attention to this, given that our conclusion was going to be that we were not making a detection claim. I thought they were wrong on that. I am proud that we moved it into the abstract, because goodness knows this was the thing we had to sweat so hard over in this search, but not just because it was hard but because it's important.

After more discussion of the kind of problems involved in calculating the exact significance of the event, this scientist stated:

> [A]t the end of the day I think it is exactly the right thing for us to hold out until we got the kind of evidence that wouldn't cause us to lose sleep overnight after this went out. So even though I'm now exposed to the world as extremely hard line on that issue compared with other people I respect, I still think it's the right thing to do.

The overall solution to these problems put forward by this scientist was to go for a level of statistical significance that would overwhelm all the unresolvable problems:

> I'm gradually coming to have more respect for the folklore that we could inherit from high-energy physics, which is "insist on 5 sigma before you claim a detection." And we've resisted that because, in part, our noise is so non-Gaussian that sigma is not necessarily a meaningful thing for us and that is probably why we haven't adopted it, but it has been put on the table as a kind of criterion. And then people gasp because the false alarm rate associated with 5 sigma in Gaussian noise is extremely small. But it does have the one virtue that it's why they don't agonize over the trials factor or anything else because they know they've got a huge trials factor thing, and I guess they've learned that it's really hard to account for that in any accurate quantitative way—there's too many ambiguities. And so you want to get to the point where no matter if somebody's wrong about the trials factor by a factor of two it doesn't change what is the, from a statistical point of view, nearly incontrovertible nature of the discovery. Whereas we are trying to be aggressive, at different levels—on the hard-ass side other people want us to tolerate more risk—OK. But I think we're in a very different regime than people who are used to, on a monthly or yearly basis, trying to decide is this a discovery or not?

This respondent is saying that, because it is a first discovery, one must be absolutely certain, and that is why a very strong statistical justification is necessary. Taking a broader perspective, however, one might say that because this is a new branch of science that has never made a confirmed detection, the first discovery will be something in which confidence rises slowly; it will at first be suggested by indication papers and then supported by steadily less tentative claims.[5] When I put this to him, however, another

5. A good model would be Joe Weber's series of increasingly confident publications in the 1960s! (See *Gravity's Shadow*, chapter 4.)

very senior scientist said: "I think that's abrogating our responsibility as scientists."

The respondent I am talking to in the long quotations above is one of the ones who pressed the argument that the Equinox Event looked too much like a glitch to be taken seriously. In a subsequent email he told me:

> [T]he two seconds surrounding the E.E. in H1 data contained several actual glitches almost identical to the E.E. itself. They were slightly weaker, but only slightly. Separate from the weak statistics of the E.E., it would have been very difficult to defend the E.E. as a detection given that we'd have to explain its appearance in a "mini-storm" of actual glitches.

In our conversation I put to him the fact that we know that signals can look like glitches. He knows the argument but, as he says, is willing to "bare his soul": "This looks like the shit in our detector, and I won't face the world saying that the thing we found looks just like the shit."

I make the point that a signal can look like shit but can still be claimed as an event because of the statistical confidence in the coincidence. He says:

> That's right, from a statistical point of view that ought to be correct. So now I'm just telling you—I'm baring my soul as an experimental physicist. . . . Noise, when we talk about it in the abstract—we always have to live with it—it is what makes measurement hard. But there's the noise that will always be with us like the poor, and then there's the failures of craftsmanship, that people should be ashamed of, that's ill behaved. That doesn't have good statistics. That means that formal significance estimates should be more suspect than they otherwise should be. . . .

I interrupt and say that this means the noise is being counted twice—once in the statistical calculation and once when you look at the signal and say it looks like noise. My respondent agrees:

> I think you're on to it. And I think I agree with that. It does count it twice, and it counts it twice because it's bad noise because it's bad noise—it's crap noise. It's noise that shouldn't be there.

There is another problem that my respondent explains prevents him accepting the position I am representing. This is that we cannot fully trust the time-slide method for estimating background: "If I had confidence that we understood all the subtleties of our background estimates—if time shifts didn't sometimes dramatically give you the wrong answer for the background—then I would agree."

Why might time slides give the wrong or inconsistent answers? There are two problems. The first is the size of the time slides that are used. There is a lower limit on the period of a time slide. Thus, the Equinox Event lasted about 1/25th of a second, but some signals can be expected to last a second or two. If the time slides are shorter than this, it could be that coincidences between bits of the *same signal* could be counted as spurious and therefore part of the background. This would incorrectly reduce the apparent significance of any real event. On the other hand, if the time slides are too long, they might compare areas of the output of one machine with areas of the other that do not represent the background noise that was found at the time of any putative signal. As it is, the convention that seems to have been adopted, though I have not been able to find out exactly why, is that comparisons should be made between the output of one instrument and the output of the other subject to a succession of 3.25 second-long shifts. People worry, however, that changing this 3.25 seconds might produce a different result for the background. The choice of length of the time slides might give rise to an arbitrariness in the measurement of the background.

There is something of still greater concern to the community, however. The Equinox Event occurred in a short section of H1 data that was untypically highly populated by glitches. If one were to separate out that section of data and subject it alone to time slides in respect of the output of the other interferometers, one would expect the apparent background to be higher because, given that there were a lot of glitches in one of the data streams, there would be a higher probability of chance matches with glitches in the other data stream. The method as it was actually used—which compares a long stretch of data with another long stretch of data—might underrepresent the background for the short glitchy stretch. This is a powerful argument for not taking the background at face value when calculating the likely improbability of the event being due to chance—it makes it look a little less like double-counting. But, again, there is no certainty about how to use this "intuition" in a quantitative way.

These points taken into account, I again present the point that the whole design of the instrument has turned on the statistics of coincidence, but my respondent returns to an earlier passage of the discussion: "[W]e go back to the point that I keep saying: I think it is the morally correct stand for someone trying to move from knowledge that we don't have to knowledge that we do have to insist on a pretty high standard of evidence."

What is a heated argument finishes with a joke. He says: "And 3 sigma things happen all the time."

I say that ignoring 3 sigma results would rule out all quantitative social science. He says that the same would apply to medical research, and we both agree that maybe that wouldn't be a bad thing.

Now the scene shifts to the Los Angeles airport, and I begin to write this book.

7 Gravity's Ghost

What was the Equinox Event? As it happens, we now know exactly what it was—a blind injection with a waveform almost identical to the waveform extracted by the Burst Group. The astrophysical event imaginatively represented by those who injected the waveform was pretty well what the Burst Group said it would have been had it been real. The Burst Group did its calculations right!

The real Equinox Event—the blind injection—is much less interesting, however, than the metaphysical Equinox Event—the ghost that haunted the collaboration from September 2007 until March 2009. The existence of "Gravity's Ghost" makes it possible to unpeel the layers of argument and inference that will surround any marginal first detection by the collaboration. There is layer upon layer. First, however, an apologia.

Apologia: The Equinox Event and the Role of the Sociologist

As things stand, a sociologist is a guest in the house of gravitational wave detection physics. Perhaps as physics moves into the twenty-first century, there will be more such guests whose role will be recognized. In my home university department, which houses half of the Centre for Economic and Social Aspects of Genomics, it is a standing joke that there are more social scientists and ethicists studying stem cells than there are biologists. Microbiology accepts that its social legitimacy depends on this phalanx of outsiders watching every move. Physics, that untypical corner of science, is still protected from this kind of scrutiny, which is why the sociologist's role remains so unusual and so delicate. In the case of gravitational wave detection there are good reasons for privacy to do with making sure unscrupulous and untrained people do not get hold of data or proto-findings before they have been thoroughly analyzed, but there seems to be no reason why privacy should continue to be the default position in physics, an enterprise so heavily dependent on taxpayers' money. It would seem more natural that the right to privacy be justified on each occasion it was invoked. Still, a guest I am.

In this chapter and the next the sociologist's role gives rise to some reflection on how things might be done differently; this might not be thought proper, coming from a guest. For example, from the sociologist's perspective it seems that there is a tension between the model of high-energy physics, which many in the community endorse, and the essentially *pioneering* science of gravitational wave detection.

In the spirit of the Anti-Forensic Principle, the subject of this sociological evaluation is not individuals and their intentions but the unfolding logic of "roles" or argumentative "positions" within the institutions of, first gravitational wave physics and, second, science as a whole. Those roles are "illustrated" by the comments made by the individuals who are quoted here, any of whom is more than capable of switching from one role to another for the purposes of argument and analysis. Likewise, the sociologist's views that color the analysis should be seen as the product of the sociologist's role, not the particular sociologist who is writing these lines.

What is, or should be, special about the sociological perspective is that it retains a distance from the day-to-day activity of science—a distance that sometimes makes it possible to reflect more easily on strain or tension whereas full-blown participants are too busy *living* such tensions to reflect on them. To reflect and analyze properly requires, first, a journey as close to the heartland of the science as possible, where the participants have all the advantages, but then, and only later, a stepping back, which is not a natural or necessary activity for a participant. But for the sociologist not to step back and try to open the window on a larger perspective, even though it could be seen to violate the etiquette of a guest, would be to fail to fulfil the duties associated with the role.

There is a moral danger here nevertheless. Though gravitational wave physics has occupied the larger part of my academic life, it is not the whole of my academic life. I do not spend the hours and hours every day of every week that the physicists spend, calculating, writing programs, analyzing data, and fixing bugs. I delete dozens of emails connected with gravitational wave physics every day without reading them; the physicists have to read them and respond to them. I attend the occasional telecon when things look interesting; the physicists attend two or three telecons a week, often at unsocial hours. I do not spend nights away from my family in remote locations working shifts on the interferometers. Gravitational wave detection is my respondents' world in a way that it is not my world. Though I put a lot of effort into observing and understanding that world, it does not compare with the effort, physical, mental, and emotional, that they put into making and living it. My reference groups are different—not high-energy physicists and astronomers but social scientists, philosophers, scientists who want to reflect on the meaning of their work, and perhaps a few general readers. Furthermore, when it comes to the algebra, the computer programs, and the calculations, I remain very much an outsider.[1]

Worse, for the time being I occupy, as Peter Saulson put it, the "bully pulpit."[2] Right now, duty or no, I am the only one who is writing a book about gravitational wave physics, and this gives me more space and opportunity to talk about it to a wide audience than the scientists have themselves. In consolation, when the discovery of gravitational waves is finally confirmed, the sociological commentary is almost certainly going to be trampled over in the triumph, a state of affairs that is nicely captured in the old joke: *Priest from pulpit*: "In the retreat the lame will be in the van." *Cynical member of congregation sotto voce*: "But not for long."

As a new science unfolds it is impossible to grasp the whole buzzing, blooming confusion of events. If there were no wrong choices, no company would ever go bankrupt, no general would ever lose a battle, and no cars, planes, or space-shuttles would ever crash. From the sociological perspective it can be seen that certain choices associated with the Equinox Event could have been made differently. But there is no culpability: it is just that the buzzing, booming confusion always intrudes in unexpected ways on the perfect world we are bound to believe we can create.

1. But on the relationship between mathematical understanding of physics and other kinds of understanding, see Collins 2007.

2. The term, coined by Theodore Roosevelt to describe the U.S. Senate, means a highly favorable platform for making a point. "Bully" was used in the original context as an adjective meaning "excellent" or "great."

The Layers of the Equinox Onion

We start to examine the layers of argument and inference surrounding Gravity's Ghost from the inside and work outwards. At the center of the science we see the still unresolved and seemingly irresolvable playing out of a set of tensions. The first tension is the freezing of protocols versus the application of common sense. The teams invent rules designed to prevent any overt or subconscious post hoc massaging of data that would lead to false statistical inferences. All development work must be done on the "playground" data or on the contrived coincidence data generated by a time slide. Only after the protocols are frozen is the "box" opened on the real data. But this procedure can fall foul of common sense, as the airplane event so dramatically demonstrates. If the frozen protocols have not anticipated everything, the unanticipated factor produces a tension between statistical propriety and the truth of the matter. It needed a vote in the case of the airplane event for the truth of the matter to triumph, but the vote was not unanimous—the tension remains unresolved; what, according the canonical model, should have been compelling and universal logic turned out to be a choice.

In any case, the idea of freezing protocols is placed under strain from the very beginning by the online searches. Common sense demands that online searches be carried out. Events loud enough to be seen prior to the application of the full panoply of refined statistical techniques have to be given special and immediate attention if only because they need to be brought quickly to the notice of astronomers working with neutrino bursts or the electromagnetic spectrum. As soon as an online search has spotted something, the clear distinction between playtime and real analysis is violated. There is no resolution except the very special vigilance that the logic of blind testing implies can never be sufficient.

The second tension, closely related to the first, is statistical purity versus craftsmanship. On the one hand, there is a stream of numbers that emerges from the extended causal chains of events triggered by the interferometers. The numbers include data quality flags which imply that we take "this" data more seriously than "that" data. In principle, a completely automated computer program operated by a trained pigeon should be able to analyze the data and, once the parameters are set, say "we cannot conclude there are gravitational waves here" or "it can be said, with the following degree of confidence, that there are gravitational waves there." In gravitational wave physics as we know it, it seems that this cannot be done. Most of the gravitational wave scientists hold the position that there is still a craft element to the work and that the data has to be looked at in concert with a close examination of the working of the machine.[3] In this case, the stretch of data which contained H1's contribution to the Equinox Event, on close examination, and only on close examination, was found to be populated by glitches that looked like the Equinox waveform. The role of the experimentalist, as opposed to the pure data analyst, gives rise to doubt about H1's contribution to the "coincidence." If this doubt gains the upper hand, there has been no coincidence. Seen from the different points of view associated with the many roles within the community, this was either the assiduous application of the experimenter's craft or the double-counting of noise driven by an overriding desire to find "reasons not to believe."

A complicating element is that experimental craftsmanship can be applied both ways. The most famous example of a positive application is Robert Millikan's analysis, in 1909, of his oil drop experiment, an experiment, incidentally, carried out at Caltech, the home base of the successive directors of LIGO. Millikan wanted to prove that the unit of electric charge was "integral"; that is, no electric charge could be subdivided beyond the unit of charge carried by the electron. To do this he needed to show that the charge on an oil drop was never less than a multiple of this unit—there were no "fractional charges." But Millikan's experimental notebooks show that he did find fractional charges—or, at least, apparent fractional charges. He applied his craft skills retrospectively to dismiss them as experimental artifacts—something that under the collaboration's protocols would be quite beyond the pale. But the judgment of history has confirmed Millikan's approach. He applied experimental craft to extract the right result where the actual data could easily have been seen to support his opponents, who said that charge was indefinitely divisible.[4] The history of science is full of similar examples; craft judgment is applied after the event to filter the data and extract a result which turns out to be right.[5] In gravitational wave detection, however, experimental craft knowledge used after the event is nearly always used to make the event—a potential result—disappear. The approach is explicit: craft skills can be used in a proper way to reduce the salience of something potentially positive like the Equinox Event; craft skills cannot be used to reduce the salience of an event when an upper

3. And only very close examination would reveal whether completely blind analysis could be done in fields such as high-energy physics, even though from the outside it appears that it could.

4. The locus classicus for the discussion of Millikan's experiment is Holton 1978, 25–83. See Franklin 1997 for an opposite view.

5. See Collins and Pinch 1998.

limit is being set, however, because this would be to use them to make the outcome more astrophysically significant. The fuss over the airplane event arose because, for once, the opposite was allowed. The community is conservative—no mistake is being made by applying craftsmanship retrospectively so long as less science is being claimed rather than more.

Craftsmanship could be applied to enhance an event by using an understanding of the machine to explain and filter out more noise. If the causes of background events can be understood, they can be ruled out of the background. The less chance coincidences, the more do remaining events stand out—the higher their statistical significance. The Bison group's result were seen by almost everyone outside of Bison's immediate group to demand rejection because his method suffered from post hoc decision-making. But, to repeat, the glitchiness around H1 was allowed to be applied post hoc because it helped to eliminate an event.[6] One can, then, see the built-in technical bias in the procedures. The bias seems reasonable, but, assiduously applied, it would have ruled out many of the great pioneering results in science.

Not unrelated to this point is the judgment about the relaxation of vetoes. Too much relaxation could look suspicious, but some relaxation is necessary, as the events at Arcadia revealed. How should that choice be made?

These tensions are elements within the larger tension between the acceptance of Type I versus Type II errors—false positives versus false negatives. The strain is at the center of every statistical science, and, as we have seen in chapter 1, it goes right back to Joe Weber and the start of the gravitational wave detection business. Contemporary gravitational wave science has demonstrated a strong leaning toward the avoidance of Type I errors and away from the risk of claiming a false positive. Fear that this mindset had become pathological was the inspiration for the blind injections.

Moving out a layer, all of the above tensions come under the larger argument over objectivity versus subjectivity in statistics—the topic of chapter 5—which has been revisited throughout the book. Starting with the particular case of gravitational wave detection, there are the uncertainties and choices over the interval, or length, of the time slides and also the uncertainty over how to handle the fact that time slides applied to

6. Ironically, the full extent and significance of this glitchiness was first spotted by Bison and his team.

short glitchy sections of data alone would show a bigger background—that is more likelihood of the Equinox Event being due to chance—than time slides applied to the whole data set. There is no clear, "mechanical" way to resolve the problem.

In chapter 5 it was shown that the meaning of any number reported at the end of a statistical process also depends on both knowing and understanding a large body of unknowable things about the history and contemporary activities of teams and individuals. It depends on what individual experimenters had been doing with the data prior to their reporting of the result—too much tuning activity now being the standard reason for explaining why Weber's results were wrong. It depends on what the members of the experimental team were thinking when they produced the number—that the Rome Group had *not* been thinking of the particular hour within the twenty-four that showed a peak prior to them finding the peak being a crucial argument against their claim to have found a marginally interesting peak. It depends on what other people in the team, and perhaps outside the team, and perhaps in private, have been doing with the data prior to the publication; and it depends not just on knowing but understanding the significance of what has been done—the seemingly irresolvable problem of how to calculate a trials factor. Lastly, it depends on what a community is willing to believe—the changing sociology of what counts as a reasonable belief in that community; this is the argument that extraordinary findings require extraordinary evidence, often referred to as "Hume's argument concerning miracles." As both Bayesians and Frequentists understand, what counts as "extraordinary" is a movable feast.

The 5 Sigma Solution and its Problems

One way to try to get round some of these problems is to *bury them* in statistical significance. Systematic errors and really large violations of statistical protocol aside, if the significance of the result is at the 5 sigma level—the level that has become the standard in the high-energy physics community, enforced by what I hear is sometimes referred to as the "5 sigma police"—then many of the conundrums and irresolvable issues discussed in the last paragraphs will be buried alive, as it were. Let there be an unresolved trials factor—it will never be enough to vitiate a confidence level equivalent to one chance in a million or so of being wrong. As a respondent put it, defending the 5 sigma standard:

> You want to get to the point where no matter if somebody's wrong about the trials factor by a factor of two, it doesn't change what is, from a statistical point of view, the nearly incontrovertible nature of the discovery.[7]

There are three problems with this approach. The first is that high-energy physics is different because beefing up the statistics is a matter of waiting long enough for more particles to be injected; it is all a matter of time.[8] Searches for individual sources are not like this. They are quintessentially unpredictable; one gets lucky or one does not, and one cannot control the source, which is the heavens.[9] Waiting for AdLIGO is one way to solve the problem, because there should then be a steady enough stream of sources to make the science look a bit like high-energy physics. However, if waiting for AdLIGO is too readily adopted as the only way to make sure no mistake is made, then Initial LIGO and Enhanced LIGO are being retrospectively redefined as machines which are much less able to make a detection than had been widely believed. iLIGO and eLIGO were built because they might discover something, and the likelihood of that discovery depends on their range. But, assuming that the first discovery is going to be of a weak event, too much caution means the calculated range effectively is cut by a factor of a few. For some, there was cause for concern even in the caution exhibited in the rejection of the Equinox Event because it occurred in a glitchy patch of data. As a very senior and very prestigious respondent put it: "And so effectively what we've done is made our sensitivity a factor of two or three less good by the act of being so conservative. And that's scary."

To go all the way up to 5 sigma, enormous luck aside, makes it much more likely that it is going to be necessary to wait for Advanced LIGO for a signal that can be talked about, and AdLIGO will not be producing

7. On the other hand, Allan Franklin (private communication) points out a case where a published 5.2 sigma result for the discovery of the "pentaquark" turned out not to be supported in subsequent experiments. See S. Stepanyan, K. Hicks, et al. 2003.

8. According to Krige (2001), Carlo Rubbia found a way of taking less time to reach a satisfactory level of statistical significance for the discovery of a new particle, and won a Nobel Prize, by drawing on the results of one of his rivals.

9. Actually, both the stochastic background and the pulsar searches could, in principle, just wait for more evidence if computer time was not so limited: the longer the observation time the more the signal, if there is one, builds up in terms of its statistics. In principle, if any vestigial signal were integrated for long enough it would become significant, though putting this fully into practice may await more sensitive instruments. One of my respondents argued that waiting for AdLIGO, given that it had been planned to be built from the outset, was actually like waiting for more particles to be generated in a high-energy physics experiment even though at first sight it seems like building a new instrument!

good data until about 2015 at the earliest. When Initial LIGO was being planned, an argument put by a critic was that there was no need to build two devices because, if there was little chance of an actual discovery, one site would be enough for all the development work to be completed until the genuine astronomical observatory was ready to be constructed. The 5 sigma level, in coming much closer to ruling out a discovery with the early devices, enhances the validity of this argument and intimates that some money could have been saved without significantly slowing progress—at least, if we lived in a world without politics or human emotions.[10] The missing entry from the "Downside" column of table 1 (see chapter 3)—on the advantages and disadvantages of blind injections—can now be filled in. The blind injections are forcing the collaboration to reveal the true working sensitivity of the detectors rather than their theoretical or measured sensitivity. In the absence of blind injections (or real events to come in S6), the difference would never have been exposed. As it is, there is a chance that we will see a continuing mismatch between the implicit promise of the early generations of LIGO detectors and the performance that they are actually permitted to achieve. For the sociologist, an upside corresponding to this is that the scientists are being forced into declaring their hands with respect to what is going to count as a detection, whereas without the blind injections the question could have been left unanswered until AdLIGO (putatively) makes the problem go away.

The third problem with the 5 sigma standard is that it may turn on a false model of gravitational wave detection physics in the years leading up to Advanced LIGO. LIGO has been a success—in the sense of it being an apparatus of near miraculous sensitivity completed in a not totally unreasonable time in the face of enormous skepticism that it could be done at all. This success was brought about under the leadership of high-energy physicists. Virgo was largely driven by former high-energy physicists too. I argue in *Gravity's Shadow* that only the high-energy physicists could understand the subtleties, including the political imperatives, and the brutal and unsubtle mechanisms of big science, necessary to bring about this success. The understanding of these subtleties and the consequent application of a degree of cognitive and managerial savagery, were a necessity if LIGO was

10. *Gravity's Shadow*, 717. All the data analysis protocols could have been developed by working with time-shifted data sets from one interferometer. I argue in *Gravity's Shadow* that two sites had to be constructed to keep senior scientists interested enough to spend their lives on the project, and this still seems correct. Even this justification becomes shaky in retrospect if such a high a standard for a detection is set that no discovery becomes possible.

ever to survive and reach a reasonable level of sensitivity. But now that LIGO is "on air," contemporary high-energy physics might be the wrong model.

At the end of my discussion about statistical significance with Jay Marx (see chapter 5, p. 99), I asked him whether 5 sigma, though it been shown to the right standard for high-energy physics, was also the right standard for gravitational wave physics:

> Collins: Do you think that this kind of standard [5 sigma] is OK for gravitational wave detection as well?
>
> Marx: I have no idea—you don't know until you've had a sample of gravitational waves and you can understand the statistical significance the analysis gives about background and how accurately information about the source can be extracted compared to what nature tells you. There's no experience yet. The comments about weak interactions in particle physics were based on years of experience by many people in that field with a significant number of experiments.

Here Marx says that knowing the right standard is a matter of experience and that there is too little of it in gravitational wave detection to gauge the statistical standard that will nearly always precipitate correct findings.

The deeper point might be, however, that pre-AdLIGO gravitational wave detection physics is not equivalent to technologically developed sciences like high-energy physics. LIGO and Virgo and the rest are right at the beginning of their scientific lives. To impose statistical confidence standards appropriate to a technologically developed science on a science that has not yet made its first, tentative, discovery could be to stifle it. Imposing these standards could have stifled high-energy physics in its early days too, as even it did not adhere to the "standards of high-energy physics" when it was first developing. As the extract from Alan Franklin's book shows (see chapter 5, p. 98), high-energy physics did not shift from using 3 sigma as its standard for a discovery until the 1970s.[11] There followed a period when 4 sigma was counted as satisfactory. Furthermore, high-energy physics admits and publishes papers, even in high prestige outlets such as *Physical Review Letters*, that are entitled "evidence for" rather than "observation of." "Evidence for" papers do not demand the 5 sigma standard—they accept

11. Franklin has found (private correspondence) at least one instance of a 2.3 standard deviation result mentioned in an earlier paper.

a lower standard.[12] The Equinox Event could have been considered as a potential candidate for "evidence for" even if it was never going to quite reach even that standard. That it was not seems, in a good part, to do with the "myth" of the "Italian" style of "indicazioni."[13]

Another science, which was stunningly successful as a pioneer of a whole new branch of physics, but which used statistical techniques that are very far from the standards of today's high-energy physics, was gravitational wave detection! Joe Weber simply reported lists of events in the late 1960s and early 1970s when he was becoming one of the most famous scientists in the world. He did not report any levels of significance measured by standard deviations until well into the 1970s and then the levels were "all over the place." The obvious response to this is to say that Weber is a prime exemplar of how to do things wrong. But this is, perhaps, too glib. First, Weber did found the whole international billion dollar science of gravitational wave detection—he was an enormous success. Second, the major complaints about Weber's techniques were not to do with the absolute levels of significance reported but the way he generated them. Once one goes in for post hoc data analysis, any level of significance can be generated—statistical significance can never compensate for systematic error. Perhaps contemporary gravitational wave detection science should be taking notice of the standards, if not methods, of its hugely successful pioneer; perhaps the wrong myth is being promulgated. Today's gravitational wave detection is Weber in the early 1960s writ large.

There is a counterargument to the relaxation of statistical standards. As explained at the beginning of chapter 4, science has changed. In many of the earlier cases, such as that of Millikan, scientists had quite a bit to go on to guide their "instinct" when it came to choosing which bits of data to keep and which to throw out, or when it was wise to go with a claim even though the statistics were poor. Nowadays, because we are looking at more and more marginal events in observational science, there is nothing to go on but statistics—there is nothing, or almost nothing, to guide the scientific instincts when a judgment has to be made about whether this or that statistically unlikely concatenation of events signifies a real

12. For example, Allan Franklin points out to me that a paper by Abe, Albrow, et al., published in 2004 in *Physical Review Letters* is entitled "Evidence for Top Quark Production in $p^{-}p$ Collisions at $\sqrt{s} = 1.8$ TeV" and offers a 2.8 sigma significance. An extended version of the paper with the same title referring to the same data was published in the same year in *Physical Review*.

13. The 2.5 sigma associated with the Equinox Event is probably not enough even for "evidence for," but the point is that the Event was never even looked at as a possible "evidence for," and it never even reached "cannot be ruled out."

event. If there is nothing to guide the instincts, so that the numbers have to stand on their own, a higher standard may be justified—and, forgetting what Weber achieved in the way of founding a field, the demand for a high standard would have applied to Weber too.[14] On the other hand, since the standard of today's high-energy physics could well rule out evidence for gravitational waves for many years, it may argue still more strongly for the use of statistics to produce "indications" and for the gradualist rather than binary approach.

As can be seen, there are two ways of looking at the question of whether gravitational wave detection should treat itself like high-energy physics in respect of statistical confidence. There are, as one might say, the "statistical experience" argument and the "nascent science" argument. The statistical experience argument holds that since the Second World War a huge amount of experience on how to do physics has been gathered and that we now know that nothing below 5 sigma is reliable if statistics are all we have to go on in making a discovery—that is what physics has taught us. The nascent science argument holds that, while this might be a valuable lesson, it applies only to the well-developed sciences because their standards will stifle new sciences. It is hard to decide which of these two positions fits the case, but there is probably something to be learned from in both. The trick is, perhaps, not to buy entirely into one to the complete exclusion of the other.

Hammering the Equinox Event into Shape

At the risk of some repetition it is worth re-describing Gravity's Ghost as an ingot of knowledge forged by pressures coming from the past, the future, and the present.

The Past

The meaning of the Event is shaped by the whole history of failed claims to have seen gravitational waves from Weber's first announcements to the Rome Group's 2002 paper. These events are treated as moments of shame

14. Weber took advantage of this argument in handling the spurious results he reported when he ran, as he thought, one of his own bars in coincidence with the detector of David Douglas at Rochester (*Gravity's Shadow*, chapter 11). He found a level of excess coincidences with a 2.6 sigma significance, but when he found that the bars were actually running out of coincidence by more than four hours, he claimed that 2.6 sigma was not significant by the standards of (mid-1970s) physics.

for physics which are not to be repeated. "Gravitational wave detection is a science for flakes" is something that many members of the collaboration have heard before and do not want to hear again.

Ironically, of course, there would be no gravitational wave detection were it not for this "flaky" history. There is now hardly a person who will not admit that without Joe Weber's crazy enterprise there would be no LIGO, no GEO, no Japanese detectors, no Australian effort, and probably no Virgo. Furthermore, there would probably be no LIGO without the widely promulgated theoretical calculations of the strengths of a variety of relatively speculative sources that, under favorable conditions, would fall within the sensitivity of LIGO—possibilities going well beyond an initial discovery of an inspiraling binary neutron star system or a supernova.[15] These calculations were easy to misread as probable outcomes; examine the small print and nothing was promised, but politicians do not read small print. Even the experimentalists' role left space for the expressed idea that something unexpected was bound to turn up, given that people were making claims to the effect that "LIGO was the first detector to reach a level of sensitivity that made it possible to see gravitational waves," and "whenever a new instrument with an order of magnitude more sensitivity points at the heavens unexpected discoveries are made, and this one has two orders of magnitude more sensitivity."

Physics runs on optimism: the optimism of gravitational wave detection science is indicated by the figures for the range of the interferometers recorded in the daily performance logs and represented in figure 2 (in chapter 1). Figure 2 shows the interferometers on a bad day, but on most days the range of L1 and L2 was reported as about 15 Mpc. That 15 Mpc did not take into account retrospective "craftwork" that could deal a body blow to a marginal signal. Once again, this is not in the "small print," and any of my respondents is in a position to claim that it cannot be *demonstrated* that there was a significant expectation that iLIGO and eLIGO could see an event. It cannot be *shown* that the caution associated with 5 sigma and the application of craft skills to a signal has effectively reduced the range of the instruments, since the range was never publicly defined in an exact way. Waiting for AdLIGO could be said to be consistent with any document in

15. Furthermore, at a number of meetings I have heard Kip Thorne describing LIGO's range in terms of is ability to see the inspiraling of massive black-hole binary systems rather than in terms of the more usual inspiraling binary neutron-stars (which imply a much shorter range), but there is no evidence that large black-hole binaries on the point of inspiral exist or that the universe is old enough to have given rise to them.

the archives. And yet the optimism associated with the new instruments was palpable. It can be illustrated, quantitatively, with the incident of the Ladbroke's bet.

In 2004, the British betting firm Ladbroke's, opened a book on whether gravitational waves would be discovered before 2010. The criterion would be a report in *New Scientist*. The odds offered were 500:1 (rumor has it that Ladbroke's was advised by one of LIGO's old enemies), but within a couple of weeks the fevered betting among the scientists and those who knew what was going on inside the project dropped the odds to 3:1 and Ladbroke's closed the book.[16] One can say from this, and from the scuttlebutt, that most of the scientist-insiders thought the fact that a signal was going to be found was worth a bet even at low odds.

As far as outsiders are concerned, optimism may be the only fuel available for a big-spending science, given that the political system goes so much faster than the establishment of new knowledge. "Give us these hundreds of millions, with these opportunity costs for the rest of your constituency, so that we might deliver a bit of knowledge long after you are out of office" works for scientists—who are ready to devote their lives to future generations of knowledge makers—but it doesn't work so well for politicians. Ironically, then, the history of failure to achieve either detections or the fulfilment of promises might also be the condition of such success as gravitational wave detection has achieved.[17]

So far so normal, but the power of the historical legacy of gravitational wave detection has been amplified by the use of the history of failure in the internal politics of the field. In the normal way, incorrect or incredible claims are given a short "run for their money" and then ignored. In gravitational wave physics, however, the failures were more salient. As soon as the big interferometers demanded one hundred times as much in the way of funding as the previous bar technology, the results coming from the much less sensitive bar detectors had to be shown to be worthless. Joe Weber forced the confrontation, were it not going to happen anyway, by writing to his congressional representative insisting that the interferometers were a waste of taxpayers money since his technique could detect gravitational waves for a fraction of the cost. Joe Weber had to be shot down in the most

16. I squeezed in relatively late with £100.00 at 6:1!

17. Pinch (1986) points out that pioneering neutrino detectors were built on the back of promises of a high flux of detectable neutrinos. The promises were successively downgraded as the detectors were first funded and then built.

explicit way or LIGO could not be justified.[18] Thereafter, each positive result reported by the bars was likely to suffer the same fate, and this may explain some of the vigor of the rejection of the 2002 paper.

There is a danger that this history of forthright rebuttals of anything that looks like "indications" has painted gravitational wave detection into a corner. To escape from the corner, it is necessary to understand the causes of the historical disdain for provisional results; that way the disdain can be to some extent discounted. The "highly respected member's" stated view could show the way out of the corner. Remember, he said of the Rome Group's 2002 paper:

> I read what they did. I think statistically they fudged things a little bit. . . . But I didn't think what they said was so bad. They said we cannot rule out that this is a gravitational wave. It has some marginal significance and we cannot rule out that it is a gravitational wave. . . . I thought that what they had was not unreasonable. And of course the first evidence that we're going to have is going to be marginal evidence. It's going to be at the edge. . . .

Even though no one now thinks the 2002 findings were right—and that includes their discoverers, since the findings were not confirmed by subsequent data—it is hard to be a pioneering science when to be provisional is to be despised.

The Future

The impact of the future is also such as to make the field—all except that part represented by its most longstanding members—risk averse.

At the September 2008 Amsterdam meeting a young but middle-ranking scientist, responding to the claim that the only way to be sure that one had seen a gravitational wave was to have an electromagnetic counterpart, said:

> The optical counterpart could be argued that's the only way forward but it's not. There's a second way forward. The second way forward is to say "if it is a gravitational wave that you're looking at, in fact, there is probably more than one per year if you look at the numbers. [This says] "maybe wait." And . . . depending on the actual run schedules, that could be right thing to do. If we

18. The history of this conflict and the evidence for its politicization is described in *Gravity's Shadow*, in the section around chapter 21.

> were going to shut down for five years before we had an instrument again, then I'd say you have no choice but do the best you can with what you have, but when you do know that you are about to go . . .

The implication was that if you know you have a new instrument coming on air soon that should see many more such events, you should wait for it.

Again, as reported, at the Arcadia meeting a respondent argued that gravitational wave physicists can wait for a high-energy physics-like buildup of data when the more advanced detectors come on line:

> We're trying to do it now, but we know that if we can't do it now, it's *not* just a matter that well we wait and the universe may or may not get it. We're expecting to have, in a reasonable amount of time, such a dramatic improvement of the event rate, that if something like this is right, the event rate should go up by a factor of a thousand [he is thinking of Advanced LIGO] and they should be falling into our laps.

Likewise, as reported above, there was a question from the floor at the opening of the envelope session where the same sentiment was expressed:

> Are we too detection oriented? In other words, why do we need to make a detection in S5 or in S6? What is the rush? . . . What is the scientific basis for rush?

Another respondent pointed out to me that the younger people in the field, who would anyway not be earning Nobel Prizes for an early discovery, had nothing to gain and everything to fear from too early an announcement: it is bad for the chances of tenure if one is associated with an incorrect claim. The period of waiting is almost certainly no longer than ten years at the time of writing (2009). Unless something is very wrong, by that time Ad-LIGO will have shaken down and the data will be pouring out with around one event a day to look at. The future, as always, favors the young.

On the other hand, the old timers have everything to gain from an early announcement—a chance to enjoy or describe the detection of gravitational waves before they become senile or die.[19] For some of the old timers there is also the chance for a Nobel Prize.

19. For the purposes of this analysis, I am an "old timer."

The Present

One force from the present is the matter of upper limits. Upper limits are the means of turning the lead of non-discovery into the gold of a story with enough significance to be published. In the early days of upper limit publication, "the story" was that the instruments had simply been made to work. More recently, as described above, some of the upper limits have gained a degree of astrophysical or cosmological significance. LIGO has been able to set a not entirely uninteresting limit on the flux of the cosmic background gravitational radiation, a more interesting upper limit on the flux emitted by the Crab Nebula pulsar, showing that it cannot be hugely asymmetrical, and an upper limit that astrophysicists found interesting, showing that the gamma-ray burst emanating from the direction of Andromeda was not an inspiraling neutron star system within the galaxy. The majority of upper limit claims, however, have no astrophysical significance.

It is not immediately obvious why prestigious journals have been ready to publish so many upper limit papers, especially papers that bear on no astrophysical phenomena of real interest. Some of these upper limits were based on data from past runs where the devices had not yet reached full sensitivity even as new data was being generated and analyzed. The plethora of upper limit papers worries some of the scientists. At a meeting in March 2008, a plan was put forward by one group to publish fifteen to twenty upper limit papers based on S5 data. A scientist who had been with LIGO a long time remarked:

> When you are publishing that many papers, when you have not seen a source, it's a little bit like we're butterfly collectors who have not found a butterfly yet. . . . I think we are going to lose credibility if [we can't make a case for each new upper limit being] way, way better . . . People will get tired of reviewing our stuff.

Even the astrophysically significant events were not being received with unqualified acclaim by outsiders, as the extract in box 1 reveals. It is from the blog of "Orbiting Frog," someone who knows what went on at the American Astronomical Society (AAS) conference in June 2008.

A more subtle question about upper limits was not, so I understand from respondents, even being asked at the more general physics meetings where upper limit results were being unveiled. Analysis of the nature of calibration (see *Gravity's Shadow,* chapters 2 and 10) indicates the possibility that, since LIGO has never seen a gravitational wave, it is impossible

My Beef with Gravity Waves

Posted on 03 June 2008

Yesterday there was brief moment when I thought that they had *announced* the *first detection* of a gravitational wave by *LIGO*. Needless to say, this turned out to not be the case. If it were then you would have heard about it - most likely from a newsreader doing their very best.

The paper that caused this trouble describes how AAS has been used to place a low-limit on some properties of the Crab Nebula pulsar (you can *read it here* if you like). The way this paper was announced at the AAS meeting in St. Louis made it sound like they had a detection. But they didn't and don't (yet).

(http://orbitingfrog.com/blog/2008/06/03/my-beef-with-gravity-waves/)

Box 1. Orbiting Frog's comment on upper limit papers

to be sure that seeing the absence of waves actually sets an upper limit; it could be that there is something wrong with the process of detection. One senior respondent indicated to me (December 2007) that to point to this alternative explanation was "not to think like a scientist."

Another respondent suggested the following more nuanced analysis:

> Your attitude questioning our calibration method isn't entirely wrong—it depends how much one wants to "bracket out" for any particular purpose. We couldn't work if we didn't think we could correctly simulate (for calibration purposes) an incident gravitational wave by moving the mirrors using magnets. But as soon as we succeed in seeing real waves, we'll claim it as an accomplishment to have validated the theory behind our having believed in our prior calibration. I think that what [. . .] was saying was that you are being impolite/anti-social in raising the question now. . . . We can always make a mistake in calibration, but the odds of us being completely wrong here are small—small enough so that, if we don't see something with Ad-LIGO and get around to questioning that aspect of our experiment, it would be interesting.

In this remark the sociology of scientific knowledge is encapsulated: it is not a matter of what is logically possible, it is what it is "polite" to ask at a particular juncture—that is, it is what is considered to be within the envelope of legitimate questions. When the first respondent said that to think that way was not to think like a scientist, he could be glossed as saying that my questioning indicated incomplete socialization into the way scientists belonging to this field are meant to think at this time. In the same sense, "the Italians" could have been said not to know how think like (interferometer) scientists when they published their 2002 paper.

To nail down the logic of the situation, consider what a flawless calibration of, say, LIGO's ability to detect inspiraling binary systems would be like. It would require one to contact, superluminally, a superhuman entity in, perhaps, the Virgo cluster, and ask it to have created and set in motion the final moments of a well-specified inspiral system at, say, a distance of exactly fifty light years exactly fifty years before 6 a.m. local time at some specified date. Then, at 6 a.m., a search could be made for its effect on LIGO.

It is "impolite," or demonstrates an inability to think like a scientist, to point out the differences between the way LIGO actually calibrates itself and this ideal way, and the potential created by the space between the ideal and the actual. For instance, inside this space between ideal and actual can be found disputes about the theory of the generation and transmission of gravitational waves and their consequent detectability by interferometers. These disputes are not salient in the community's discussions—and to acquire the "interactional expertise" pertaining to the community is to know that these disputes are not worthy of discussion—but they have not entirely gone away; they have just spun out into the fringes of the field. For example, a paper claiming that LIGO and other interferometers are based on a flawed theory of the transmission of gravitational waves was promulgated on arXiv, the physics electronic preprint server, as recently as 2008, and this paper was based on previous work of long standing, albeit discussion of that work within the community has died.[20] These claims do not have to be correct to have a life inside the logical space between ideal and actual calibration. The possible gap in our understanding of what happens between the generation and detection of gravitational waves is not closed by calibration, it is assumed to be closed. And as my second respondent pointed out, this gap is acknowledged, in that the theory will of how they work will be said to have been validated when gravitational waves are discovered. At that point the theory will no longer be something that is simply taken for granted; it will be treated as having been validated and worthy of a triumph.

20. I am grateful to Dan Kennefick for bringing this material to my attention: Fred I. Cooperstock, "Energy Localization in General Relativity: A New Hypothesis," *Foundations of Physics* 22 (1992): 1011–24; Luis Bel, "Static Elastic Deformations in General Relativity," electronic preprint gr-qc/9609045 (1996) from the archive http://xxx.lanl.gov; R. Aldrovandi, J. G. Pereira, Roldao da Rocha, and H. K. Vu, "Nonlinear Gravitational Waves: Their Form and Effects," arXiv:0809.2911v1 (2008).

The second kind of gap between ideal calibration and calibration as it is carried out can be seen once we accept the assumption that the theories of generation and transmission are correct. In that case a gravitational wave is encountered by an interferometer as a ripple in space-time that squeezes it one way while stretching it the other and then reverses the cycle. But a ripple in space-time impacts upon the entire earth and every part of the interferometer including not only the mirrors but the mirror mounts, and it affects every part simultaneously. For many years there were disputes about whether gravitational waves could be detected in principle since, as one might say, the ruler one uses to measure the changing separation of the parts of the detector is affected just as much as the instrument itself.[21] Theory now assures us that gravitational wave detectors can work, though the dispute about them does not seem to have been closed until the 1960s, and then only because Joe Weber actually built a detector and people started to argue about whether he had seen anything rather than whether he *could* see anything in principle.

That an interferometer is believed to be able to detect gravitational waves is a matter of a theory that relates the movement of its solid parts—the hanging mirrors—to the effect on the light that bounces between them. It is not a trivial theory. It may be virtually inconceivable to most interferometer scientists that the theory could be wrong, but the *calibration* does not prove that the theory is right. When a gravitational wave hits the interferometer, the signal is meant to be read off as the force required to hold the mirrors still in spite of the gravitational wave's attempts to move them. But an injection of energy for the purposes of testing the collaboration's ability to detect a real wave is made by putting a signal into the coils that press on the magnets fixed to just one of the mirrors on just one of the arms, and, because of the theoretically deduced differential response of the mirrors to waves of different frequencies, quite a bit of calculation goes into determining what this signal should be. There is no reason to think this will not imitate the effect of a wave on the arms, and it may, once more, be almost inconceivable that it won't, but when gravity waves are detected, that all these inferences were correct will still count as a triumph because calibration does not test them, it assumes them.[22] In this passage I am being im-

21. These disputes are discussed in Kennefick 2007.

22. Within the collaboration the term "calibration" is used for something a little different—the rather more simple calibrating of the various individual components of the interferometer. It is from aggregating the results from all these individual tests that the range of the interferometers, as illustrated in figure 2, is built up. What I am referring to as calibration, the members of the calibra-

polite; it is impolite to talk of such things in the current state of the art—it is a sign of defective socialization, like bad behavior at a dinner party.

Actually, if one goes a little further outside the community, one can discover the questions being asked by others. They are being asked by people who seem to be physicists but are unschooled in the etiquette of gravitational wave detection. What can be found in the physics blogs is illustrated by the remarks in box 2, which refer to the announcement of the upper limit on the Crab Nebula pulsar's spin-down energy.

That these considerations tend to float around on the fringe of the community and no longer enter its heart are the real sociological point. The critics—those bitter critics from the astronomical community, so active in the early 1990s, who were trying to stop LIGO being funded—have gone to ground.[23] Now that the political battle over LIGO's funding had been lost by the critics, they have nothing to gain from belittling LIGO's achievements. That no one deploys the potential of "the calibration criticism" shows how successful LIGO has been at building credibility.

This achievement of credibility means that upper limits can be announced and published with confidence in the absence of fear about the calibration question. But there is a downside. It means the younger segment of the collaboration can build a career on publishing risk-free upper limits and do not have to worry about risky positive discoveries. There is nothing in it for the old timers, but, ironically, the very success of LIGO—its credibility and legitimacy within the scientific community—is one of the indirect forces pushing toward a conservative interpretation of things like the Equinox Event—exactly the opposite of the forces experienced by the dying bar-detector community in the 1990s and early 2000s as the giant LIGO drew closer.[24] One can only wonder how LIGO and Virgo data analysis would feel today if the approaching advanced LIGO technology belonged to a rival group.

---

tion would talk of as "hardware injections" (thanks to Mike Landry for helping me through this thicket). My use of "calibration" accords with the more "philosophical" meaning.

23. See *Gravity's Shadow*, chapter 27.

24. Still more faux credibility can be generated when journalists report an upper limit results as a positive finding. The following headline from a journalist's blog glosses the result as a positive statement that the Crab Nebula pulsar is emitting gravitational waves:

> Crab Nebula Pulsar Leaks Energy through Gravitational Waves
> Up to 4 percent of the radiated energy is converted into gravitational waves
> (http://news.softpedia.com/news/Crab-Nebula-Pulsar-Leaks-Energy-Through-Gravitational-Waves-87171.shtml)

jimmy boo June 06, 2008 @ 05:06AM

This could be a good opportunity to test the gravitational wave hypothesis. If a method could be devised (perhaps by looking for irregularities/precessing of the radio pulse) to determine the shape of the pulsar, the expected gravity wave radiation from this could be calculated. If the mass distribution is sufficiently non-spherical such that the expected radiation is above the detection limit of LIGO it would provide a strong suggestion that 1) there is something wrong with LIGO, or (and more interestingly) 2) there is something wrong with our understanding/belief in the existence of gravitational waves ...

(http://episteme.arstechnica.com/eve/personal?x_myspace_page=profile&u=8530045777 31)

Posted by *Iztaru* 06/02/08 15:23
Rank: 5/5 after 2 votes

They say the lack of gravitational waves gives clues about the structure of the pulsar. But that is assuming the GWs exist and this pulsar has not produced them.

Their analysis of the structure of the pulsar is void until someone detects an actual gravitational wave.

I remember a while ago they were very exited because they didn't detected any GWs from a gamma ray burst coming from a nearby galaxy because that ruled-out several scenarios for the source of the burst. But again, it is assuming GWs exist.

Is there anyone making analysis of these scenarios but assuming GWs do not exist? I think that is a must, until an actual wave can be detected.

(http://www.physorg.com/news131629044.html)

Box 2. Two examples from physics blogs

The other force from the present has already been discussed. It is the scientists' peer groups in related areas of physics and astronomy working with the canonical model of science, and/or working in established, technologically well-developed, fields. The binary model of discovery, and the 5 sigma imperative, are also imported via the recent experience, in high-energy physics—the paradigmatic big science—of many of the collaboration's members. If the nascent science argument has some truth in it, so that the statistical experience argument does not hold the field unopposed, it might be better to think of LIGO as a small science in big science clothing. Gravitational wave detection was a small science in the days of Joe Weber and even the cryogenic bars. To build LIGO and the other interferometers took the political, economic, and organizational techniques of big science. But now that the interferometers are collecting their first data, it may have become, effectively, a small science once more. Big sciences are characterized by advances of around an order of magnitude in sensitivity that are justified by predictions of what will be found based on the past successes of the previous generations. There are always surprises, but there is some order to the progress. But there is not so much as one past success in terrestrial gravitational wave detection on which to build experience, and the increase in sensitivity of the detectors has been two or three orders of magnitude. This kind of risk—this degree of floating free of the past with no lifeline of past success—is normally the prerogative of small sciences. To build a science one needs results, and, if every one of these has to be perfect, maybe the new shoot of a science will wither and die before it reaches the full glare of sunlight above the canopy of its competitors. *Horribile dictu*, it may be best for the big-but-really-small science of gravitational wave detection to act like "the Italians"—at least in some respects. These respects include endorsing "evidential collectivism" rather than "evidential individualism"—exposing provisional findings to the wider scientific community rather than keeping every disagreement or uncertainty behind closed doors. Remember that at the Equinox Event discussion in Arcadia we heard the following:

> . . . for the most part we are almost certainly going to be in a case where we are not going to have the kind of confidence that some of us would perhaps like to have, and I think we need to get used to the idea that we may have to, as a group, say we have seen something and put ourselves on the line over it.
>
> I think we have to decide are we willing to live with the possibility of not seeing something that is really there in order to be conservative or not.

Conclusion to the Main Part of the Book

That the Equinox Event was an artifact made some difference to the way it was analyzed by the scientists. Some analysts guessed what it was and put less effort into trying to show whether or not it was noise than they might otherwise have done. Indeed, the whole blind injections exercise illustrates the difficulties of "social engineering": an effort that was supposed to change scientists' mindset away from rejecting potential discoveries to embracing them had the opposite effect on at least some and sowed confusion among some others. Overall, however, the exercise seems to have been a scientific success because so much was learned from it, as the Postscript below will show.

Sociologically, the blind injections have been a huge success. Gravity's Ghost was almost as good as a genuine liminal event in the way it revealed both the true meaning of what it means to discover something in the "exact sciences" and also the tensions and judgments that surround the process even in an esoteric field of physics. In this it continues the work of Wave Two sociology of science and continues the application of it to gravitational wave detection, which is exemplified by much of *Gravity's Shadow*. In this particular case something new has been done. The observation of the proto-discovery process has made it possible to explore the day-to-day meaning of statistical analysis of data in a depth that has not been done before so far as I know.[25] There is nothing missing from the account of such a statistics-based discovery except the still more burning problems that might be associated with a full-blown announcement and, of course, with the process of reception in the wider scientific community. Statistical analysis has been shown to be a social process in a very concrete way. Most of that was summarized in the first part of this chapter, which was organized on the principle of "stripping away layers."

Discovery and statistical analysis being social processes, it has been possible, in the second part of the chapter—the one employing the metaphor of the ingot of knowledge—to consider the social and cognitive forces that press upon the meaning of the Event. Throughout, the underlying idea is that what is happening in a social group is revealed by what its members take seriously enough to talk about and what they reject without consciously thinking about it. Thus it has been possible to see something like the "calibration question" ceasing to have a place in debate within the core of the science though it lives on in logic and can still be found to be discussed among the poorly socialized, such as those at the fringes of the field (and those taking the role of the poorly socialized—namely myself). I have tried to show how what is considered seriously and what comprises meaning is affected by pushes and pulls from all sides. Sometimes these forces will be almost palpable. The force of history is an example in this field, and it has been seen how its power is maintained by the recounting of myths—stories about historical incidents which are taken to embody vital lessons.

25. Which is not to say that there is no sociology of statistics: see for example Mackenzie 1981, which analyzes the social roots of the correlation coefficient and its basis in questions of race.

Less powerful influences have also been discussed in the final section. These include the existence of "escape routes" circumventing the felt need to make immediate discoveries, such as upper limit publications and the prospect of more powerful detectors to come. It is impossible to say how strong these less powerful influences are, or even if they make much difference at all, but the point is that the close examination of the data analysis process reveals how they could be having an effect. The analysis shows how any such effect would be mediated by the discourse of the field. Snatches of that discourse even suggest that these considerations are playing a role.

The book goes beyond Wave Two, however, in that it tentatively explores the possibility that the reflective sociological stance might contribute to the discourse as well as record it. It has been suggested that from the distanced position it appears that there is a tension between gravitational wave detection as a pioneering science and gravitational wave detection as a big well-established science. With great trepidation it has been suggested that clarifying this matter might help resolve some of the tensions in the statistical analysis caused by, for example, the desire to meet the 5 sigma criterion while not having the means to do it with the current generations of detectors.[26]

In the next section of the book the distance from the field is going to be suddenly increased—to use a metaphor from cosmology, a kind of "sociological inflation" is going to take place. That is why there is no chapter 8 but rather something that I have called an "envoi." An envoi, according to the *Chamber's Dictionary*, "is the concluding part of a poem or a book; the author's final words, especially the short stanza concluding a poem written in certain archaic metrical forms."

26. Allan Franklin points out that some of the processes and problems described here do have precedents in high-energy physics. Franklin is preparing a paper, which he kindly refers to informally as a "footnote" to this book, which will show where similar problems are to be found in the recent history of high-energy physics. The paper will also show how the problems were debated and resolved in those cases.

ENVOI
Science in the Twenty-First Century

This book has dealt with the problems of detection of liminal events. But maybe this is not the future of gravitational wave detection. Maybe there is some big surprise or a colossal stroke of astrophysical luck to come. A big close event may be visible in many different ways via electromagnetic or other effects. With the detection of such an event, the science would be propelled from provider of upper limits to almost incontrovertible certainty-maker, without the in-between stage of "indications."

From a sociological viewpoint, it would be a pity if the interferometers had the stroke of scientific luck that would suddenly put an end to the debates and uncertainties that have lasted for nearly fifty years. There is more to be learned sociologically from the dilemmas surrounding marginal events and from a gradual phase-transition from nonbelief to belief than from dramatic discoveries. That is one reason for writing this book now, before the arguments can be overtaken by a strong event of this sort. The arguments will survive such an eventuality, but gravitational wave physics would no longer exemplify them so nicely, and the agony of the Equinox Event would be quickly forgotten. In sum, the sociology of knowledge is best served by an uncertain science. There is, however, a bigger question. The topic now shifts from gravitational wave detection in particular to science as a whole and to its role in social life. Here gravitational wave detection becomes merely an illustrative example rather than the central topic.

Science as a Reflection of Society

Science no longer has the unquestioned authority it had when it was winning wars and promising power too cheap to meter. Nowadays it is beset by raids on its epistemological grounding from academe, a backlash from religion, an attack on its professionalism from free-market ideology, and scorn from those who think a simple life without technology is the only salvation for the human race. The attack on experts and expertise that has come with the academic movement known as postmodernism actually does the same work as the attack on the professions begun by Margaret Thatcher and Ronald Reagan and takes the same view on experts as religious fundamentalists. The academic movement holds that expertise has no epistemological warrant, the political movement holds that the judgment of professionals has to be replaced by quasi-markets in which every aspect of performance is measured and compared so that it can be properly priced, and the fundamentalists hold that expertise is worthless in the face of revelation. The arms of this grotesque pincer are squeezing science from three sides.

But is science worth preserving? What is science? Nowadays it sometimes appears that science has no unique cultural identity left. Like a teenager, today's science is continually borrowing from others' cultural repertoires to gain attention. We have science-as-Hollywood with its superstars, their vanity bolstered by the media industry and a new style of popular science publishing. We have science as a religion, with Stephen Hawking and Charles Darwin as its figureheads. Hawking sells millions of copies of a completely incomprehensible text which ordinary people revere as they might once have revered the Latin Bible. Meanwhile the media treat Hawking and his mystic utterances as revelation, while other physicists see, or are said to see, the "Face of God" in the skies. Richard Dawkins and his colleagues attack organized religions with the gospel of Darwin, borrowing religion's rhetoric, beatification, and iconography. There is also science as the slick player in the market, from Silicon Valley to the muscular-capitalism start-ups of the new biology and its instant millionaires. If this is science in the twenty-first century, then there is nothing to learn and nothing to preserve, at least not in the way of values. As far as cultural values are concerned, this kind of science is all reflection and no substance. Such science cannot claim to be a source of transcendent value, since there is nothing transcendent about it. This way, even without the pincer, science will garrotte itself as a cultural movement.[1]

The pressures to act this way are, of course, enormous. Economic pressures lead governments to demand visible short-term economic pay-offs from the public funders of science. Big science has to compete for funds in the political arena, where publicity is a dominant force. The massively costly astronomy and particle physics need the front pages linking them to the story of our beginnings and our ultimate fate—previously the agenda of religion. Individual scientists find a ready justification for pushing their noses into the trough because to glamorize is to survive.[2]

Maybe this is the fate of science—to be a secular religion servicing the economy and the entertainment industry. But science, at least as exemplified by the story of gravitational wave detection, has the potential to lead not just follow. It has the potential to provide an object lesson in how to make good judgments in a society beset by technological dilemmas. For more than three hundred years the old-fashioned values of science have seeped into Western societies like the air we breath. Imagine a society without any place at all for the cultural authority of science. It would have surrendered all responsibility to politics, market forces, or competing modes of revelation, and it would be a dystopia—at least as anyone who prefers reason to force would see it. This is not to say that science is the only cultural institution the removal of which would lead to a dystopia, but it is central to the kind of society most of us prefer to live in.

The Central Values of Science as a Cultural Institution

It is hard to list the special values of science, because it is an activity only vaguely defined by the "family resemblances" of its different parts.[3] One

1. For what appears to be an alternative perspective on contemporary capitalism-linked science, see Shapin 2008. Shapin makes no distinction between derivative and essential values and appears to treat with equanimity a science-capitalism nexus which, while blind to matters of race, gives differentially favorable treatment to young, physically fit, and competitive persons of local origin.

2. The political maneuverings of scientists have been wonderfully documented over the years by Dan Greenberg (e.g., 2001).

3. The term "family resemblance" was introduced by the philosopher Ludwig Wittgenstein, who argued that the idea of "game" has no set of clear defining principles but that games were linked by family resemblance—which can be thought of as a set of overlapping sets, each containing a subset of game-like qualities but whose extremes need have little or nothing in common; games, after all, run all the way from professional football to not stepping on the cracks in the

can make progress, however, by imaginatively taking away different elements and seeing if what is left can still be called science. Thus, one can take away the ability to see the face of god and still have something recognizable as science; one can take away the best-selling books that no-one understands and still have science; one can take away the religious war against religion and still have science; one can take away the venture capitalists and still have science; and one can take away the front page stories and the superstars and still have science. These features of science as a cultural institution are merely "derivative."

One the other hand, one cannot take away integrity in the search for evidence and honesty in declaring one's results and still have science; one cannot take away a willingness to listen to anyone's scientific theories and findings irrespective of race, creed, or social eccentricity and still have science; one cannot take away the readiness to expose one's findings to criticism and debate and still have science; one cannot take away the idea that the best theories will be able to specify the means by which they could be shown to be wrong and still have science; one cannot take away the idea that a lone voice might be right while all the rest are wrong and still have science; one cannot take away the idea that good experimentation or theorization usually demand high levels of craft skills and still have science; and one cannot take away the idea that, in virtue of their experience, some are more capable than others at both producing scientific knowledge and at criticizing it and still have science. These features of science are "essential," not derivative.[4]

Three qualifications are in order. First, not every one of these values belongs to science alone. For example, most religious institutions, every professional institution except those to do with magic, illusion, and crime, and indeed every society as whole, must value honesty if they are to last. But honesty seems more logically integral to science than to the others. Everywhere else honesty must be the default or average position; in science it seems to be always vital or science simply is not being done. Therefore science provides a kind of "sea-anchor" for the value of honesty and integrity.

---

sidewalk. (Whether there is "little" or "nothing" in common is not so clear; here I am trying to identify some things that all, or nearly all, science has in common.)

4. The essential/derivative distinction is radically at variance with the popular "Machiavellian" approaches to science found in the field of Science and Technology Studies, which, laying all the stress on detailed observation of science as it is encountered, see the achievement of scientific success as essentially a political process.

Second, not every value listed in the "essential" category is equally central. There is still an argument going on about whether the theory of the historical evolution of species satisfied the falsifiability tenet. But even if it does not, we know that the overall theory has been strengthened by small-scale demonstrations of the mechanisms in the laboratory and that these do satisfy the tenet. Thus, overall science needs the falsifiability value even if there are violations here and there.

Third, generalizing the second qualification, honesty and integrity aside, one can still have passages of scientific research that produce valuable findings where one or more of the values is violated.[5] The list of values, contrary to the way they have sometimes been thought of, does not comprise a set of necessary *conditions* for the production of good scientific work.[6] Rather, the overall "institution" of science is formed by these values, and it follows that the day-to-day "form-of-life" of science is built up through actions driven by the "formative intentions" of scientists who are guided by these values. The idea of family resemblance shows how such a set of values can hang together to form a cultural institution without always needing to be manifest in every one of its corresponding activities all the time.[7]

Given these qualifications, it could be said that the essential values of science are far more important than science's substantive products and findings. The essential values of science are worth preserving so that they can continue to seep out into society as a whole and help to shape the way we live our lives. That is why the pincer needs to be resisted even if, in the short term, the maintenance of the conditions for continued production of scientific substance and findings seem to rest with the derivative values. Either one sees that this is the case and chooses the central values or one does not; if one cannot see these values as good in themselves and prefers not to choose them, argument ceases and force prevails.[8] As far as I have been able to observe, the heartland of gravitational wave detection science exemplifies these values quite closely.

5. As the Second Wave of science studies has clearly demonstrated.

6. Arguably, Merton's (1942) "norms of science," which overlap with the values listed here, were initially justified as a set of conditions for the efficient functioning of science.

7. One of the persistent errors of some critical approaches to science is to take it that discovering lapses from the central values of science—something that is easily done—shows that there are no central values.

8. The choice to make the essential scientific values (not findings) central to the life of a society can be called "elective modernism" (see Collins 2009).

Another thing that cannot be taken away and still leave science—another essential value—is the idea of the replicability of findings; replicability is a corollary of the idea that there is anything stable to investigate. Replication often cannot be carried through in practice, as when unique events are observed or apparatuses are so costly that only one can be built, but the idea of replicability still informs what is going on. Thus, should there have been only a unique observation, the idea is that anyone who was in a position to see it would have seen the same thing. If there is only one apparatus, the idea is that another similar apparatus would make the same discoveries or that anyone else who knew how to run the machine would find the same things.[9] Adherence to a value does not always mean that the value has to be instantiated there and then, but it does mean that the value remains an aspiration even when circumstances cannot allow it to be realized.

Including replicability, another consequence of the essential values of science is that secure claims cannot be based on the authority of individuals or other unique sources. Holy men or women cannot "reveal" the truth in virtue of their unique relationship with a deity. If the relationship is unique, then it cannot be replicated (not even in principle) nor criticized, so it cannot belong to science. The same goes for books of obscure origins or with obscure authorship. What is written in such a book cannot command scientific respect in virtue of the fact that there is something special about it as an object: its contents have to be open to criticism and investigation. Inter alia, these principles rule out creationism as a part of science, because the story of the creation depends too heavily on a particular book of obscure origin and its interpreters. These principles also rule out the more technical version of creationism, "intelligent design," because it too is heavily dependent on a book of obscure origin for its ideas and because it seems impossible in principle to think of any falsifying observation or experiment—that is, one that would show that any still unexplained state of affairs could *not* be counted as the work of an intelligent creator.

Another thing that follows from these principles is that works of science should be as clear as it is possible to make them, failing destructive simplification. The simpler and better the explanation, the easier to criticize; to make a work obscure beyond necessity is to make it, effectively, more private, and privacy is incompatible with freedom to criticize. Note that at the other end of the spectrum of valuable cultural institutions, creating

9. Replicability is, as one might say, a "philosophical" idea. To see how it works in practice, see Collins [1985] 1992.

works that are open to competing interpretations and debate by those that consume them is an essential feature of the arts.[10]

The principles also indicate why the imitation of major features of non-scientific cultures can be antithetical to science—the means destroying the ends. If millions of people are being encouraged to treasure, for their scientific content, books that are so far beyond their ability to criticize as to be completely incomprehensible, then they are being encouraged to think of scientific worth as like religious worth—based on the authority of the author or the text. The same goes for all "glamorization" of science and scientists: as soon as the person, rather than the ideas and findings, becomes important, then the very idea of science is being subverted. It is the same when the virtues of science are advertised for their commercial potential. The balance of scientific worth is not the same as the balance of commercial worth. That science generates more economic value than it costs is probably true, but that science is worth keeping primarily because of its value to the economy is not true.

Science as an Object Lesson in Judgment

Contrary to what some commentators claim, quantitative exactness is another thing that *can* be taken away from science without destroying it. It is a shame that the idea that exact quantitative analysis is crucial to science is so widespread because it does great harm. For example, a social-scientific finding has little chance of influencing government policies unless it is expressed in quantitative terms, whereas many quantitative social science results, when they are not simply wrong, are of no social significance. In social science qualitative findings are often far more robust and repeatable than quantitative findings and often more socially significant, yet they make small headway with policy-makers. As has been seen in chapter 5, even in the physical sciences the expression of a result in statistical terms is often the tip of an iceberg of hidden judgments, assumptions, and choices, and yet, to those who consume rather than produce them, the numbers still have a force quite out of proportion to their true meaning.

Certainty, along with the binary model of discovery and the Nobel Prize, is another thing that can be taken away while leaving science intact.

10. The argument is worked out in detail in chapter 5 of Collins and Evans 2007, under the heading of "Locus of Legitimate Interpretation" (LLI). If the LLI is forced pathologically close to the producer in the arts and humanities, it is "scientism." If forced pathologically close to the consumer in the sciences and technologies, it is "artism."

Indeed, one might say that certainty is the province of revelation rather than science, and I have heard even potential Nobel laureates claiming that lust for prizes distorts and damages science. Even a philosopher of science such as Popper can point out that all scientific claims are essentially provisional.

A shallower but more important point about the false allure of certainty was made in the Introduction: most science is applied to such messy domains that good judgment rather than certain calculation is quite obviously the best that can be done. If certainty and quantitative exactness were the keys to science, then science would be restricted to that small corner described at the outset as comprising the products of Newton, Einstein, quantum theory, and so on—the so-called "exact sciences." But there is a much larger domain of "inexact sciences," which impact much more immediately on our lives. This is the domain where society needs its object lesson on how to make decisions as well as how to live its life.

Of course, gravitational wave detection fits naturally into the realm of the "exact sciences"; its precursors are Newton and Einstein, and, as far as its place in the spectrum of sciences is concerned, it takes its "reference group" to be high-energy physics and the like. I suspect that when the final announcement of the terrestrial detection of gravitational waves is confirmed, it will be impossible for the scientists to resist the retrospective redescription of their field as one of heroic point-discovery, and I suspect they will be unable to resist the glorification of the enterprise in culturally derivative ways. At that point, as I have suggested, "the lame" (that is, the gradualist and uncertain description of discovery) will be trampled under the stampede; the "Drums of Heaven" will be said to have sounded and "Einstein's Unfinished Symphony" will be said to have been completed.[11] Perhaps more important to the mass of working physicists who have dedicated their lives to the project will be the right to face their critics from the other exact sciences with pride; making the most of the newly accomplished similarity between gravitational wave detection and the sciences of their peers will be irresistible.

But what we have seen both here and in much preceding work involving detailed analysis of the sciences is that even the exact sciences are inexact when examined under the microscope. Here we see that this inexact

11. With apologies to David Blair, whose (so far as I know, unpublished) manuscript on gravitational wave detection was given the first title, and to Marcia Bartusiak, whose very good, popular introduction to the field was given the second.

face is visible even to the metaphorical "naked eye" when the science is engaged in true pioneering. The analysis of the Equinox Event, though the event was an artifact, was a moment of true pioneering as far as the procedures of science are concerned. As we have seen, judgment after judgment had to be made. Indeed, as *Gravity's Shadow* reveals, the entire multi-decade history of gravitational wave detection has consisted of judgment after judgment.

I want to suggest that the mythical hero of the science of gravitational wave detection should be not Einstein but Galileo. Einstein may have conceived of the idea of gravitational waves, but Galileo represents science as a leader of social understanding—in Galileo's case, the end of the geocentric universe. I have already suggested that the essential scientific values encountered in the heartland of gravitational wave science are a self-evidently good model for social conduct. What I am now suggesting is that the *inexact* science of gravitational wave detection (not the triumphalist and *exact* version almost certain to come) is an excellent model for judgment wherever technology and society intersect in an inexact way.

The science from which we can learn to live our technological lives is hard, frustrating science full of flawed judgments which are nevertheless the best judgments there can be. Barring massive luck, the first indication of gravitational waves will be "the best guess about what this funny looking blip in the statistics might mean." Such a display of how to understand the world and what it is to understand it as well as possible is a vital thing that science has to offer society. It is something that belongs to science alone, something which allows it to lead, not something derived or reflected.

One of the senior scientists in LIGO is quoted above as saying that to offer only gradualist and tentative results in the search for gravitational waves would "abrogate our responsibility as scientists." He was reflecting a very widespread view among the collaboration's members as at the beginning of 2009. From the sociological and political perspective offered here, the opposite is the case. Scientists' responsibility lies in making the best possible technical judgments, not in *revealing* the truth. To represent every judgment as a calculated certainty is to abrogate *social* responsibility. To be a producer of certainties is, at best, to consign oneself to the nonexemplary sciences—the corner of the scientific world that has dominated, and distorted, Western thought with examples of what it claims to be a perfect and, worse, *attainable* mode of knowledge-making. To strive too much for certainty is to abrogate the responsibility of taking that leading role in Western societies that only science as a cultural activity can assume.

If scientific judgment is not applied to the uncertain world that faces us, the role of finding solutions will be relegated to populism, fundamentalism, force, or what amounts to the same thing, the market. If we want to maintain the society we live in, the technical element of judgments had better continue to be based on technical experience. By exemplifying scientific values, and judgment under uncertainty, gravitational wave physics could be among the sciences that play a still more important role in twenty-first century political and social life.

The view put forward in this last section of the book is uncompromisingly sociological—perhaps a violation of etiquette characteristic of the improperly socialized. One of its bizarre implications is that the more incorrect results physics makes public the better, so long as they are meant to display the process of a gradual move toward better understanding and better judgments. It follows that all the rejected incorrect or partial claims described in *Gravity's Shadow* could be more valuable in social terms than the point-discovery and even the gravitational astronomy to come. The trouble is that the physicists are ashamed of all those incorrect results. But why should they be? They all arise from physicists doing their job with a burning passion. It is likely that the shame is generated, not by a disappointed public, but by close rivals in the fight for resources—one set of physicists damning another in their attempt to get a share of their funds. That kind of competition is not an essential value of science—scientists can get things wrong and science very definitely still goes on. In the science in the twenty-first century, more valuable than the truth is the demonstration, and the guidance this can give to society, of how experts with integrity judge when they do not know.

POSTSCRIPT
Thinking after Arcadia

Dates can be important in contemporaneous work. The first draft of this book was written in about five weeks. The first key was struck in Los Angeles airport on 19 March 2009, and, after a week or so of primping after completion, a confidential copy of the manuscript was sent on 12 May to my gravitational-wave physicist friend, Peter Saulson, for his comments.[1] The manuscript has undergone continual polishing since then, partly in response to the remarks of those mentioned in the Acknowledgements, while the Envoi gained a life of its own. But up to and including chapter 7, the *substance* of what appears here is very close to the draft that was completed as April turned into May in 2009. Those chapters represent the best I could do in the way of recording physicists' thinking and talking from the first intimations of the Equinox Event to the end of the Arcadia meeting in the middle of March 2009.

Things started moving very fast after that, and this Postscript is an attempt to capture the way the blind injection experience affected things in the few months that followed. There were two further large meetings, one in Orsay and one in Budapest, held in June and September respectively. I did not attend either of these meetings but accessed them by other means: being by now a kind

1. The five weeks of composing was, of course, based on a store of knowledge and experience built up over nearly forty years, on some preexisting draft papers on statistics, and in particular on hundreds of pages of notes referring to the blind injections.

of "honorary" member of the LIGO Scientific Collaboration, and having been given the passwords that allow me access to almost everything that is put up on the web by members of the community, I was able to view the slides presented at the meetings and read emails exchanged in response to them; I was also able to access some of the discussion in Budapest via a telephone link, and colleagues reported additional goings on.

It is hardly worth mentioning that the Inspiral Group agreed that they must find a way to move faster and do all the checks on the data as soon as possible. Once more, there is nothing more of sociological significance to say about their response. Of greater potential sociological significance were the reactions to the Equinox Event. The task that confronted those who wished to be in a position to make a claim about the detection of a burst of gravitational waves from statistical likelihood alone—without the reassurance of correlated optical effects or a well-defined inspiral waveform—was being reassessed. In particular, some were now suggesting that, if events demanded it, "evidence for" claims with a much lower level of statistical significance could come before "observation of" claims. The notion of what could be presented in the journals was becoming less binary and more "Italian." Nevertheless, the divide between the Virgo and LIGO teams was, if anything, becoming sharper, with the Virgo people in general pressing a much harder line. But even one member of that group, who had previously been of the conservative persuasion, displayed slides at both meetings which included the following sentiments:

> **What if we have to deal with a single candidate without any counterpart?**
> Seems hard to reach the "discovery" criterion.
> The "evidence" criterion is not out of reach.
> It is likely that our first detection paper title will be "Evidence for . . ."
> Then "Strong evidence . . ." etc. . . . as we are able to increase the significance of candidates.
> Continuum of possibilities between first evidence and a clear and convincing detection.
> First paper likely to be "*Evidence for blabla . . .*"
> A long (and painful) (and exciting) road in front of us.

Though these remarks show that something less than discovery was now being countenanced by one or two Virgo people, a hard line in respect of discovery itself was still being pushed by everyone from Virgo whose remarks or presentations were accessible to me. Most were in fact pressing an even harder line, which was being described as the "Virgo position"—a position, incidentally, that is consistent with the account provided by 'S' (see chapter 4).[2] This Virgo position was that nothing at all could be announced about a single event unless it (a) correlated with optical effects or (b) demonstrated a well-defined waveform. A third possibility was that (c) statistics alone would be sufficient so long as there were a series of repeated events—effectively a call to wait for AdLIGO. The criterion of coincidence between separated detectors—the detection principle that has informed gravitational wave detection from the days of Weber—was simply no longer to be counted as good enough on its own.

Some LIGO people agreed with this thinking, but others suggested that the caution in respect of statistical analysis that was being argued for—leaning over backwards to take into account all possible contributions to the trials factor, assuming that the non-Gaussian nature of the data made any extrapolation from the actual false alarm rate generated from time slides impossible, and refusing to accept a single event until similar events had been seen—took the gravitational wave community well beyond what was normal practice in astrophysics and perhaps even of high-energy physics itself. A specially interesting intervention came from Jay Marx, the executive director of LIGO. He represented the detection of gravitational waves as normal science, given the theory, the astrophysics, and their indirect observation by Hulse and Taylor via the slight decay in the orbit of a binary system observed over many years (see *Gravity's Shadow*). Marx suggested that 3 or 4 sigma should be enough to support an *observation* claim because an observation of this kind was to be expected.

My reading of the overall state of discussion at Budapest was that there was still considerable confusion and disagreement, sometimes exhibited by what looked like contradictory statements within individual talks. In particular, a couple of talks from Virgo spokespersons insisted, on the one hand, that no discovery could be made with a single candidate and, on the other, that whatever was found might have to be reported (just as the "Italians" argued). The apparent contradiction is made clear in the two sentiments expressed one after another on the same slide presented by a senior figure in Virgo:

2. Some LIGO scientists were surprised that they could be such things as "Virgo positions" or "LIGO positions" in a unified team, questioning once more the exact nature of the LSC-Virgo collaboration.

> We cannot afford "Evidence for . . ."
> WARNING: If the situation is borderline with respect to this we might be forced to publish our data WITHOUT claiming a detection.

The resolution of this apparent contradiction is that the putative publication flagged by the "WARNING" would discuss all the possible sources of noise in the detector that might have led to a coincidence that, though statistically significant, could not be presented even as "evidence for" in the absence of other kinds of confirmation.[3]

Pressed further by people from LIGO about the apparent contradiction in the Virgo position, another senior Virgo spokesperson, himself, on the face of it, a hard-liner, said:

> I think that this is a very difficult moment we are passing through because it is difficult to find the first event and so we may have to take a chance in accepting such an event. It is very difficult to say what is the right thing to do. Honestly, I don't have the answer. I am very worried that we might publish something that is not a real event. The solution could be repeatability, but that's also difficult because we are looking for rare events. I don't know what to do in this case—honestly. [Edited for English style by HMC]

The debate, and the dilemmas it embodies, can be summarized along three dimensions:

Dimension 1: What kind of science is gravitational wave detection?

This question has already been discussed at length in the main body of the book. GW continually compares itself with high-energy physics, but that standard may be too high for a pioneering science. In any case, the standards of high-energy physics were more relaxed when high-energy physics was developing. Furthermore, GW may be setting itself standards even higher than HEP when it tries to take into account every imaginable aspect of the trials factor problem and when it refuses to countenance the calculation of a background which goes beyond what can be directly measured with time slides. It may also be going beyond other sciences in the way it applies craft understanding of the instruments in an asymmetrical way.

3. Private communication with the author of the slide (October 2009). The author added in this private communication: "I know several cases in other fields where publication of unexplained behavior occurred, ending into nothing. Measurements continued and the effect didn't show up again."

Finally, it may be astrophysics that is the better model for the discipline; astrophysics, so at least one commentator said, operates with less conservative standards than high-energy physics.

Dimension 2: "Evidence for" or discovery claims only?

Again, this dimension has been much discussed in the body of this book. At the Orsay meeting there seemed to be a new willingness to countenance "evidence for" (something that, for what it is worth, is argued for throughout this book). But at the Budapest meeting some very strong sentiments were expressed that ran against this idea. But can "evidence for" claims be avoided if some marginal event is found in S6? We do not know what will happen in S6, but we can try a thought experiment. Imagine that four "events" are seen in S6, all of which have around a 3.5 sigma significance level. Imagine that when the envelope, or envelopes, is/are opened, three of these events are found to correspond to blind injections but the fourth correlates with nothing. Would it be credible to say nothing about the fourth event? Would a refusal to say "this looks like evidence for" demonstrate integrity or lack of integrity?

Dimension 3: Setting the rules a priori or responding to events?

For years, members of the community have looked for ways to set detection criteria in advance. The counter-position has also been argued. In Budapest both positions were once more advanced. The debate was nicely expressed in the concluding remarks of a senior member of the LIGO team:

> I've tried and I think every time I've looked at it I've failed. And that is to do what you guys have been trying to do for the last hour-and-a-half. What are the criteria you would apply, given our current state, [given] the fact that we have many different inputs—multiple detectors, external triggers, waveforms that have been calculated? How do you take that whole mix and make something that's sensible out of it? Can you make a set of rules? And that was what I heard—people making rules about how they were going to approach this. I don't think we should do it that way. I think we should make a set of concepts . . . write down a set of concepts as guides. . . . I know of no other way to go at this. . . . The Detection Committee, if they ever get exercised, will look at each case on its own merits. I don't know any other way.

Again, for what it is worth, I have argued in this book that the last sentiment is the correct one. While it is worthwhile trying to imagine what criteria one might apply in the future, these exercises are best looked at as a kind of "training" in thinking about the many possible scenarios that might arise rather than as a way of developing a rigid set of rules. The reason for saying this is set out above (in chapter 2), and it is the deep philosophical point made by Ludwig Wittgenstein that rules do not contain the rules for the own application. Mostly, the rules for application of most of the rules we use in day-to-day life are so deeply sunk in the sediment of ordinary practices and cultures that we do not even notice that we are performing a philosophically miraculous act of interpretation every time we use a rule. In cases like the one discussed here, however—the true pioneering sciences or in any radically new practice or culture—the problem of the proper application of rules becomes more salient. In such cases the only way to know how to apply a rule—for it is a certainty that every eventuality cannot be foreseen—is to work it out as events force the arguments to unfold.[4] The outcome of these lived arguments sinks down through the ocean of history to form the sediment of society—in this case, scientific society. Calculation cannot produce that sediment; only the cycle of culturally innovating life can create it. It is possible to make some attempt to rehearse the future, as war-games rehearse battles, but, as generals continually rediscover, war games are not war.

Conclusion

The discussions that took place in the post-Arcadia meetings demonstrate that the blind-injection exercise was a success. It was not a success in the way it was intended to be: it did not result in the injections being identified as pseudo-gravitational waves, and it did not give rise to a complete exercising of the detection procedures. But one may be sure that the approach to the next round of data analysis will be influenced by the impetus of blind injections much more than the last round was affected. The success is being realized one experimental run later than expected.

4. The "airplane event" illustrates this deep point of principle in almost comic form.

APPENDIX 1
The Burst Group Checklist as of October 2007

Burst Detection Checklist for GPS 874465554 event
The LSC-VIRGO Burst Working Group
October 1, 2007

1. Zero-level sanity Convert GPS times to calendar times and check for suspiciousness.
    Sep 22 2007 03:05:40 UTC not suspicious

2. Zero-level sanity
    Read e-logs for the times/days in question.
    Was there anything anomalous going on?
    seismic activity and loud/long transient at H1; nothing at L1

3. Zero-level sanity
    Check state vector of all instruments around the time of the candidate.
    Data correctly flagged?
    FOM shows all IFO's in science mode

4. Zero-level sanity
    Identify inspiral range of the instruments in order to set the scale of sensitivity. Is this typical/low/high?
    Good H1/L1 range, typical H2 range. Range drop 6 min after

5. Zero-level sanity
    Identify nearest in time hardware injections (also type/

amplitude of them). Were there ongoing stochastic/pulsar ones? When did they start and what was their amplitude?
No nearby hardware burst injection, pulsar on

6. Zero-level sanity
   How close to segment start/end for all instruments does this event occur?
   Science segment boundary is over 10,000s away for all instruments

7. Data integrity
   Check for undocumented/unauthorized/spontaneous hardware injections.
   Nothing in injection log. Nothing unexpected in any excitation channel

8. Data integrity
   Examine all possible test points recorded in frames or saved on disk. The latter part might be time critical if data are overwritten.

9. Data integrity
   Establish if there was any data tampering. O[s' responsibility]

10. Data integrity
    Check integrity of frames; check raw/RDS/DMT frames.
    data-valid flags are OK on the L1 RDSs for both observatories
    no CRC mismatch over range [874463712, 874467392)

11. The event
    Run Q-scan on RDSs/full frames and on all available instruments in the network of detectors.
    Some mildly interesting things (see link).
    H1 is very glitchy across IFO channels, must understand
    H2 is at inconsistent Q of 23, while 4km IFO's have Q of 5

12. The event
    Run Event Display on RDSs/full frames and on all available instruments in the network of detectors.
    no obvious glitches. some spikes in AS-AC,REFL-I/Q about 1 secs later.

13. The event
    What is the overall time-frequency volume of the event in each detector? Does it look consistent?
    minimal uncertainty, looks very consistent

14. The event
    What is the expected background for such a candidate?
    What is the significance of the observation given the background?
    Compare background estimated from time-slides and from first-principle Poisson estimates. Is it consistent?
    background depends on algorithm/search. at the predetermined thresholds
    for the online search we have a significance of about .03 for KW+CP.
    However the event is well above the set thresholds. This is for the probability of observing N events on Sept 21, 2007

15. The event
    How robust is the background estimate? Do randomly-chosen shifts as opposed to fixed shifts—is the result consistent? How about other steps?
    currently the background is only estimated for Sept 21

16. The event
    How robust is the significance of the event to the threshold chosen? Do a "stair-case" analysis (varying threshold) to appreciate this.
    Naturally the significance of the event depends on the threshold
    For KW+CP it varies from .03 (fg 3, bg .68) to .01 (1, .01) depending on Gamma threshold chosen (5 or 9.5)

17. The event
    If more than 2 IFOs are involved in the event, would any 2-IFO pair be able to identify it as well and with what background/significance? **easily visible in H1L1.**
    probably visible in H1H2 Q analysis (guessing from qscan)

18. The event
    If only 2-IFOs are involved, why it was not detected by the rest of the detector network?
    below KW threshold in H2, but clear signal in all 3 IFO's
    G1 and V1 too poor in sensitivity at 100 Hz

19. The event
    Examine the status of the detectors not involved in the event and establish why this is so.
    H1H2L1G1V1 all in Science Mode (wow)

20. The event

    Examine frequency content out to the KHz range—is there anything there? Will a broader bandwidth search pick it up?

    There is nothing in h(t) up to 8192 Hz. However the high frequency excess in AS_Q_0FSR for both H1 and L1 visible in the qscans should be understood (aliasing at nyquist?)

21. The event

    Run parameter estimation code.

22. The event

    Identify the sky ring/sky patch using non-coherent methods, i.e., signal timing and amplitude O['s responsibility]

23. The event

    Compare source reconstruction between coherent and incoherent methods

24. The event

    What fraction of the network's acceptance comes from this direction?

25. The event

    How robust is the candidate against the stride size/start of the analysis?

    Seen in KW and QPipeline, so it should be robust

26. The event

    How the peak times established by the method, parameter estimation and the various coherent methods compare? Is this as expected?

    using QPipeline as a reference, KW peak is 9ms later. Both algorithms give dt=4ms between H1 and L1. CorrPower peak is at −38ms with a 50ms window

27. The event

    Could it be an artifact of the prefiltering?

    Q Pipeline and CorrPower use zero-phase filtering. The positive offset of the KW trigger is due to an uncorrected filter delay

28. The event

    What is the detection efficiency for the search method calculated near to the candidate event? average/low/high w/r/t the S5?

29. The event
    Could there be any effect from lines not filtered enough and/or any other artifact? Could it be violin mode excitations? Any other mechanical resonances sneaking in?
    does not look like a line in qscan. short duration (few ms)

30. The event
    How stationary were the instruments around the time of the event?
    Quantify this both in terms of singles counting and PSD.

31. Vetoes
    What could have caused this event other than astrophysics? blind injection.
    faulty injection. common low-freq glitch

32. Vetoes
    Are there obvious environmental disturbances in the Q-scan/Event Displays?
    nothing obvious in the initial RDS qscans. however, we might want to look at qscans which display all channels regardless of significant content immediately around event time. There is a nearby low freq glitch in H0:PEM-BSC3_ACCX at −200ms, and noise in H0:PEM-BSC6_MAGZ from −6 to 1s about event time

33. Vetoes
    Are there obvious interferometric disturbances in the Q-scan/Event Displays?
    H1 shows a variety of IFO channels glitching nearby stopping 1s after event time. The glitches are not directly in coincidence with the event, but should be understood

34. Vetoes
    Examine what known earthquakes occurred around the time of the event.
    fairly quiet around time of the event

35. Vetoes
    Contact power companies and obtain known power line transients around the time of the event.

36. Vetoes
    If available, check in/out records for trucks and heavy equipment to the Lab that might not have been recorded in the ilog. Check for overpassing airplanes (airport flight logs).

37. Vetoes

Check for any switching of major electrical equipment around the time of the event that might not have been recorded in the ilog.

38. Vetoes

What are the KleineWelle (KW) triggers in auxiliary channels around the time of the candidate?

Nothing much immediately around event time (ms within .714/.718)

Glitches within the same second are weak. BSC3_ACCX is missing in H0 as KW triggers are only generated from 10–512 Hz.

39. Vetoes

If there are any overlaps with KW trigger from auxiliary channels, what is the expected background of such coincidence and what is the significance of that channel as a veto channel?

40. Vetoes

Which of the overlapping channels are safe, which are not? Analyze most recent hardware injections.

41. Vetoes

For PEM/AUX channels with measured transferred functions, is the signal present in the them consistent with the one in the GW one?

42. Vetoes

If nothing in the non-GW channels in the RDS, proceed with scanning full frames and repeat above checks.

43. Vetoes

Any known data quality flags overlapping with event? How is this dependent on DQ flag thresholds? What is the coincidence significance?

None of the DQ flags that have been evaluated so far overlap, but many flags have not yet been evaluated for this part of the run

44. Vetoes

Examine minute trends/Z-glitch/glitch-mon data.

45. Coherent Analysis

Run the H1-H2 Q analysis; anything in the H1-H2 null stream?

H1-H2 stream has norm energy of 7. H2 has energy of 15.
So H1 and H2 seem more or less consistent.

46. Coherent Analysis
    Run the r-statistic cross-correlation over all detector pairs (involved or not in the trigger); how significant each is?

47. Coherent Analysis
    Run coherent analysis/null stream burst analysis on the available detector network.
    RIDGE pipeline finds maximum statistic at location of event

48. Coherent Analysis
    Run the inspiral multi-detector coherent analysis on the available detector network and compare to the burst one.

49. Coherent analysis
    What is the best fit waveform extracted from the data?

50. Other methods
    Do other burst ETGs find the event(s)? If yes, compare extracted event parameters, including background/significance.
    The event is seen in all 3 IFO's using QPipeline, however no coincidence is done, nor thresholds set for a detection
    Event also seen in Block-Normal in H1 and L1 only
    Event seen in CWB at FAR 1/598 days over S5A.

51. Other methods
    What is the outcome of the Inspiral and/or Ringdown search around the time of the burst event? If something is present, what is the background/significance?
    No online BNS event (1–3 Msun)
    Frequency range suggests BH mass objects

52. Calibrations
    What is the calibration constants and errors around this event?

53. Calibrations
    How robust is the event analysis against calibration version?
    KW uses uncalibrated data. May affect CorrPower

54. Calibrations
    Could there be any calibration artifact?
    such a strong signal is not present in the excitation channel

55. Calibrations

Is the event identified when analysis is run on ADC data? Compare findings of an analysis starting with ADC(t) vs h(t).

event is found on DARM_ERR with KW and h(t) with QP/CP

56. Miscellanea

Check timing system of the instruments (well in sync?).

57. Miscellanea

Check for any recent reboots, software updates/reloads.

Any suspicious acquisition software changes?

58. Miscellanea

Check recent logins to the various acquisitions computers.

59. Other GW detectors

Any signature in the non LSC-VIRGO detectors? TAMA/bars online?

60. Other GW detectors

What is the expected signal size given what we know for the event?

undetectable in V1, G1 and anything else due to low freq (100Hz)

61. Non-GW detectors

Any known or "sub-prime" event in E/M or particle detectors around the globe? O['s responsibility]

62. Astrophysics

Any known sources overlapping the ring/patch on the sky corresponding to the direction of the candidate event?

63. Astrophysics

Examine events (other than the candidate) reconstructed at the same direction. Perform a directional search; if a point source is behind this, more, lower SNR events might be in our data.

64. Astrophysics

How the extracted waveforms compare to astrophysical waveforms?

What is the energy scale going into GW, assuming galactic distances?

65. Vetoes
    Create a hardware injection starting with signal waveforms corresponding to the best fit waveforms extracted from the instruments.

66. Other methods
    Take the extracted waveform per IFO and run matched filtered search in order to establish how often the specific morphologies appear in the data.

67. The event
    Run the Q-event display

68. The event
    Run the Coherent-Event-Display (CED)

69. Vetoes
    Play audio files corresponding to GW, H1+−H2, auxiliary channels

70. Vetoes
    Check that signal is the same in all photodiodes.

71. Vetoes
    Check wind speeds
    Wind speeds normal and less than 15 mph

72. Vetoes
    Check for fluctuations in power levels of TCS laser
    Normal: no mode hops or big jumps in TCS power levels

73. Vetoes
    Do a seismic Q-scan
    Some seismic noise in L0:EY_SEISY and H0:EY_SEISZ need further study.

APPENDIX 2
The Burst Group Abstract Prepared for the Arcadia Meeting

We present the results from a search for unmodelled gravitational-wave bursts in the data collected by the network of LIGO, GEO 600 and Virgo detectors between November 2006 and November 2007. Data collected when two or more out of the four LIGO/Virgo detectors were operating simultaneously is analyzed, except for a few combinations which would contribute little observation time. The total observation time analyzed is approximately 248 days. The search is performed by three different analyses and over the entire sensitive band of the instruments of 64–6000 Hz. All analysis cuts, including veto conditions, are established in a blind way using time-shifted (background) data. The overall sensitivity of the search to incoming gravitational-wave bursts expressed in terms of their root-sum-square (rss) strain amplitude hrss lies in the range of 6 × 10−22 – 6 × 10−21 Hz−1/2 [tentative] and reflects the most sensitive search for gravitational-wave bursts performed so far. One event in one of the analyses survives all selection cuts, with a strength that is marginally significant compared to the distribution of background events, and is subjected to additional investigations. Given the significance and its resemblance in frequency and waveform to background events, we do not identify this event as a gravitational-wave signal. We interpret this search result in terms of a frequentist upper limit on the rate of gravitational-wave events detectable by the instruments. When combined with the previous search using earlier (2005–2006) data from the fifth science run (S5) of the LIGO detectors, this is at the level of 3.3 events per year [tentative] at 90% confidence level. Assuming several types of plausible burst waveforms we also present event rate versus strength exclusion curves.

ACKNOWLEDGMENTS

My approach to the analysis of science is to try to understand as much of it as I can—something I call "trying to develop interactional expertise" (Collins and Evans 2007). I do my best to put this technical understanding into my analysis. In the case of the science I describe here, the statistical analysis of data, I have also had to depend on the scientists to help me out quite a lot on the technical details by checking bits of "proto-text." As always, my friend Peter Saulson, author of the best treatise on interferometric gravitational wave detection, one-time spokesperson of the LIGO scientific collaboration, and the most honest and honorable person I know, has been my constant guide in the writing of this book. That we disagree quite heartily in respect of some of its more sociological analyses and conclusions only demonstrates still further the selflessness of his contribution. I have also had help in understanding certain specific technical concepts and procedures from Alan Weinstein, Sergey Klimenko, and Mike Landry, to each of whom I sent short passages of the draft for comment. My, one time, bitter academic enemy, and now valued academic colleague, Allan Franklin, has generously looked into the history of high-energy physics in ways useful for this book but which I was technically not equipped to do for myself, and his findings are used and acknowledged in the text. Franklin intends to publish more on the recent history of high-energy physics so as to make clear where precedents and previous discussion of the kind reported here can be found. Graham Woan looked over the statistics chapter from the point of view of a gravitational wave physicist especially interested in statistics, while Deidre McCloskey looked at it from the point of view of an expert in their use in the social sciences. Robert Evans, my colleague from Cardiff, read the whole manuscript carefully and pointed out a number of places where the typical reader from science and technology studies could easily misunderstand what I was trying to say. This led me to put in more explicit pointers aimed at that community. Richard Allen, worked assiduously at copyediting to smooth the text and remove infelicities. More diffuse thanks go to the entire community of gravitational wave physicists and to my colleagues and department at Cardiff for providing an environment in which work like this can be done. Needless to say, all remaining faults are my own.

As always, Christie Henry, my editor at University of Chicago Press, has encouraged me from the outset, passing the reins for this particular volume to Karen Darling. I cannot say how wonderful it is to have found a publisher who believes you should publish what you want rather than what they want; it was a lucky day for me when I got together with Chicago.

The fieldwork was supported by a small grant from the UK Economic and Social Research Council (ESRC): "The Sociology of Discovery" (2007–2009: RES-000-22-2384).

REFERENCES

Astone, P., G. D'Agostini, and S. D'Antonio. 2003. "Bayesian Model Comparison Applied to the Explorer–Nautilus 2001 Coincidence Data." *Classical and Quantum Gravity* 20 (17): S769–S784

Astone, P., D. Babusci, M. Bassan, P. Bonifazi, P. Carelli, G. Cavallari, E. Coccia, et al. 2002. "Study of the Coincidences between the Gravitational Wave Detectors EXPLORER and NAUTILUS in 2001." *Classical and Quantum Gravity* 19 (7): 5449–65.

Astone, P., D. Babusci, M. Bassan, P. Bonifazi, P. Carelli, G. Cavallari, E. Coccia, et al. 2003. "On the Coincidence Excess Observed by the Explorer and Nautilus Gravitational Wave Detectors in the Year 2001." http://arxiv.org/archive/gr-qc/0304004.

Collins, Harry. 1985. *Changing Order: Replication and Induction in Scientific Practice.* Beverley Hills and London: Sage. 2d ed., 1992, Chicago: University of Chicago Press.

Collins, Harry. 2004. *Gravity's Shadow: The Search for Gravitational Waves*, Chicago: University of Chicago Press.

Collins, Harry. 2007. "Mathematical Understanding and the Physical Sciences." In "Case Studies of Expertise and Experience," ed. Harry Collins, special issue, *Studies in History and Philosophy of Science* 38 (4): 667–85.

Collins, Harry. 2009. "We Cannot Live by Scepticism Alone." *Nature* 458 (March): 30–31.

Collins, Harry, and Robert Evans. 2002. "The Third Wave of Science Studies: Studies of Expertise and Experience." *Social Studies of Science* 32 (2):235–96.

Collins, Harry, and Robert Evans. 2007. *Rethinking Expertise*. Chicago: University of Chicago Press.

Collins, Harry, and Robert Evans. 2008. "You Cannot be Serious! Public Understanding of Technology with Special Reference to 'Hawk-Eye.'" *Public Understanding of Science* 17 (3): 283–308. DOI 10.1177/0963662508093370.

Collins, Harry, and Martin Kusch. 1998. *The Shape of Actions: What Humans and Machines Can Do.* Cambridge, Mass: MIT Press.

Collins, Harry, and Trevor Pinch. 1998 [1993]. *The Golem: What You Should Know About Science.* Cambridge and New York: Cambridge University Press.

Collins, Harry, and Gary Sanders. 2007., "They Give You the Keys and Say 'Drive It': Managers, Referred Expertise, and Other Expertises." In "Case Studies of Expertise and Experience," ed. Harry Collins, special issue, *Studies in History and Philosophy of Science* 38 (4): 621–41.

Finn, L. S. 2003. "No Statistical Excess in EXPLORER/NAUTILUS Observations in the Year 2001." *Classical and Quantum Gravity* 20: L37–L44.

Franklin, Allan, 1997. "Millikan's Oil-Drop Experiments." *The Chemical Educator* 2:1–14.

Franklin, Allan. 1990. *Experiment, Right or Wrong.* Cambridge: Cambridge University Press.

Franklin, Allan. 2004. "Doing Much About Nothing." *Archive for History of Exact Sciences* 58: 323–79.

Greenberg, Daniel, S. 2001. *Science, Money and Politics: Political Triumph and Ethical Erosion.* Chicago: University of Chicago Press.

Holton, Gerald. 1978. *The Scientific Imagination.* Cambridge: Cambridge University Press.

Kennefick, Dan. 2007. *Traveling at the Speed of Thought: Einstein and the Quest for Gravitational Waves.* Princeton: Princeton University Press.

Krige, John. 2001. "Distrust and Discovery: The Case of the Heavy Bosons at CERN." *ISIS* 95:517–40.

Mackenzie, D. 1981. *Statistics in Britain, 1865–1930.* Edinburgh: University of Edinburgh Press.

Merton, Robert K. 1942. "Science and Technology in a Democratic Order." *Journal of Legal and Political Sociology* 1:115–26.

Mortara, J. L., I. Ahmad, et al. 1993. "Evidence Against a 17 keV Neutrino from 35S Beta Decay." *Physical Review Letters* 70:394–97.

Pinch, Trevor J. 1980. "The Three-Sigma Enigma." Paper presented to the ISA-PAREX Research Committee Meeting, Burg Deutschlandsberg, Austria, September 26–29, 1980.

Pinch, Trevor J. 1986. *Confronting Nature: The Sociology of Solar-Neutrino Detection.* Dordrecht: Reidel.

Shapin, Steven. 2008. *The Scientific Life: A Moral History of a Late Modern Vocation.* Chicago: University of Chicago Press.

Stepanyan, S., K. Hicks, et al. 2003. "Observation of an Exotic S + +1 Baryon in Exclusive Photoproduction from the Deuteron," *Physical Review Letters* 91:252001-1–252001-5.

Wittgenstein, Ludwig, 1953, *Philosophical Investigations.* Oxford: Blackwell.

Wright-Mills, C. 1940 "Situated Actions and Vocabularies of Motive." *American Sociological Review* 5 (December): 13, 904–9.

INDEX